THE FIRST CHAPTERS OF OUR COSMIC HISTORY WITH JWST

IAU SYMPOSIUM 391

COVER ILLUSTRATION:

This image from NASA's James Webb Space Telescope reveals 50,000 near-infrared light sources, including Pandora's Cluster, a massive galaxy megacluster acting as a gravitational lens. This natural magnification highlights distant red galaxies, many from the early universe, stretched into arcs by the warped fabric of spacetime. By peering into the dawn of existence, Webb brings us closer to understanding our origins and humanity's place in the vast cosmic story.

Credits: NASA, ESA, CSA, Ivo Labbe (Swinburne), Rachel Bezanson (University of Pittsburgh), Alyssa Pagan (STScI)

Link: https://webbtelescope.org/contents/media/images/2023/107/01GQQF4KP3GNVB12G6R0V8KSGM

IAU SYMPOSIUM PROCEEDINGS SERIES

INTERNATIONAL ASTRONOMICAL UNION
UNION ASTRONOMIQUE INTERNATIONALE

International Astronomical Union

IAU

THE FIRST CHAPTERS OF OUR COSMIC HISTORY WITH JWST

PROCEEDINGS OF THE 391st SYMPOSIUM OF THE INTERNATIONAL ASTRONOMICAL UNION CAPE TOWN, SOUTH AFRICA 6–9 AUGUST 2024

Edited by

THEMIYA NANAYAKKARA
Swinburne University of Technology, Australia

and

MICHAEL MASEDA
University of Wisconsin in Madison, United States of America

CAMBRIDGE
UNIVERSITY PRESS

CAMBRIDGE UNIVERSITY PRESS
University Printing House, Cambridge CB2 8BS, United Kingdom
1 Liberty Plaza, Floor 20, New York, NY 10006, USA
10 Stamford Road, Oakleigh, Melbourne 3166, Australia

First published 2025

Printed in Great Britain by Henry Ling Limited, The Dorset Press, Dorchester, DT1 1HD

Typeset in System LaTeX 2ε

A catalogue record for this book is available from the British Library Library of Congress Cataloguing in Publication data

ISBN 9781009546027 hardback
ISSN 1743-9213

Table of Contents

Preface

After decades in waiting, in 2022 JWST opened up a new window to our cosmos. Prior to JWST, we knew so much about the endpoint of the story of our Universe (i.e., the local Universe). However, we knew relatively little about the early stages of the Universe even though facilities such as the Hubble Space Telescope were pushed to their limits. Detailed studies of topics such as the first stars and galaxies, the formation and buildup of metals and dust, and the origin of the first black holes were tantalizingly close, but JWST is the facility that puts them firmly in our grasp given its excellent sensitivity, angular resolution, and wavelength coverage. The renewed infrared view of our Universe enabled by this facility has already started transforming our view and understanding of the early cosmic processes.

Scientifically, there are a number of major questions about galaxies that JWST observations have raised to date. Why are some galaxies at high redshift so luminous? Can the current ΛCDM framework accommodate such massive systems? What is the nature of "little red dots" and how do they potentially reflect the growth of supermassive black holes early in cosmic time? What causes galaxies to stop forming stars, and at what time do those processes start? All of these questions have been motivated by observations, many of which were unexpected prior to JWST, but all of them have theoretical underpinnings and rely on ongoing dialogues between observers and theorists/simulators to solve the dialogues we started or continued during the symposium.

Also, very importantly, JWST takes on the role of being the *"Observatory of the World"*. Anyone in the world without geographic limitations can request access to JWST. Proposals are only judged on scientific merit and all data products are made available publicly by the Space Telescope Science Institute (STScI). Thus, researchers from any nation, including those where funding to access large astronomical infrastructure is limited, have the opportunity to lead their own programs and be a part of the active research community using this groundbreaking space facility. The 2024 General Assembly (GA) of the International Astronomical Union (IAU), being the very first one on the African continent in the 100+ years of IAU history, allowed us to leverage this open access policy of JWST research to inspire the next generation of astronomers in Africa.

Therefore, as part of this symposium, we organized a hands-on workshop dedicated to accessing and analyzing JWST data, focusing on scientists based in Africa. Drawing from the list of invited speakers and organizers and building on the success of STScI's "master classes," 40 students and junior researchers were able to learn from the leading experts on JWST's capabilities and were shown how to utilize state-of-the-art analysis tools. Funding from the Australian High Commission in South Africa also allowed us to support 22 of the researchers to attend the GA, and to fund a networking dinner to establish working relationships between the presenters and attendees.

During the three days of the symposium, we hosted 9 invited review talks, 6 invited highlight talks, 60 total oral contributions, and 17 virtual posters.

In the spirit of the first IAU GA in Africa and the unique, fleeting moments we have with this wonderful telescope, we hope that all participants were able to benefit from the symposium and left with new science ideas and new collaborators.

Themiya Nanayakkara and Michael Maseda, co-chairs SOC,
Melbourne, Australia, 11th of December, 2024

Figure 1. Some snippets from IAUS#391: The top left panel shows the plenary talk by Dr. Nancy Levenson, while the top right panel features Prof. Richard Ellis with participants of the JWST networking dinner for African researchers. The middle panel shows the JWST hands-on workshop for African researchers, including an exchange of TimTams, an Australian delicacy shared among participants. The lower left panel shows the meeting of the Australian Acting High Commissioner Hon. David Geyer with Prof. Karl Glazebrook (PI of the JWST Australian Data Centre), hands-on workshop participants, and symposium organizers. Finally, the lower right panel shows delicacies from Franschhoek enjoyed by the organizers.

Editors

Themiya Nanayakkara
Swinburne University of Technology, Australia.

Michael Maseda
University of Wisconsin in Madison, United States of America.

Scientific Organizing Committee

Denis Burgarella (IAU Div J liaison): Laboratoire d'Astrophysique de Marseille (LAM), Aix-Marseille Université, France

Valentino González: Universidad de Chile, Chile

Kathryn Grasha: Research School of Astronomy & Astrophysics, Australian National University, Australia

Hanae Inami: Hiroshima University (広島大学), Japan

Taysun Kimm: Department of Astronomy, Yonsei University (연세대학교), South Korea

Michael Maseda (co-chair): University of Wisconsin-Madison, USA

Themiya Nanayakkara (co-chair): Centre for Astrophysics and Supercomputing, Swinburne University of Technology, Australia

Irene Shivaei: Centro de Astrobiología (CAB), Spain.

Laurence Tresse (IAU Div J liaison): Centre de Recherche Astrophysique de Lyon (CRAL), France

List of Participants

Not needed for GA symposia.

The First Chapters of Our Cosmic History with JWST
Proceedings IAU Symposium No. 391, 2025
T. Nanayakkara & M. Maseda, eds.
doi:10.1017/S1743921324002503

Advancing Scientific Discovery with JWST

N. A. Levenson

Space Telescope Science Institute
email: nlevenson@stsci.edu

Abstract. The tremendous new capabilities of JWST already enable great discoveries across broad areas of astrophysics, spanning scales from exoplanets to cosmology. We have also found some puzzling results that require re-examination of our prior assumptions and are a rich source for future investigations. JWST is remarkably complex, and the mission is possible because of the work of thousands of people around the world. Despite the complexity, it remains an open mission, with opportunity for researchers at all career stages, across a variety of institutions, and around the world. We hope to apply the lessons of our experiences to future ambitious astrophysics projects that we will pursue together.

Keywords. Early Universe, Galaxies, Stellar Physics, Exoplanets, Inclusive Astronomy

1. Introduction

The global scientific community has been making effective use of JWST since the first scientific data became available in July 2022. The sensitivity of the large (6.5-m) telescope is generally advantageous, while the infrared optimization and stability are particularly effective in key research areas, including studies of distant galaxies and exoplanets. The pace of discovery is rapid, and this contribution offers merely a snapshot of some compelling, recent results.

The observatory was decades in the making, involving an estimated 20,000 people around the world. Gardner et al. (2023) provide an overview of the observatory and its capabilities. Rigby et al. (2023) characterize the scientific performance, which is excellent. The four instruments together offer 17 different modes of use. More information about each is available in dedicated papers on the Near-Infrared Camera (NIRCam; Rieke et al. (2023)); Near-Infrared Spectrograph (NIRSpec; Böker et al. (2023)); Near-Infrared Imager and Slitless Spectrograph (NIRISS; Doyon et al. (2023)); and Mid-Infrared Instrument (MIRI; Wright et al. (2023)).

2. The Early Universe

Some of the most interesting new results using JWST examine the early Universe, which are the immediate focus of other contributions in this volume. Numerous galaxies are detected at redshift $z > 6$ and even $z > 10$, exceeding many previous predictions. The galaxies are thus brighter than anticipated, and none are primordial; *i.e.*, generations of stellar processing has already occurred. While the galaxy number counts exceed many predictions, none of the robust results are in conflict with current fundamental models of the Universe, including Λ-cold dark matter cosmology. Quantitative analysis of luminosity functions and reliable measurements of stellar masses are required to draw specific conclusions about galaxy formation and evolution, although current indications suggest

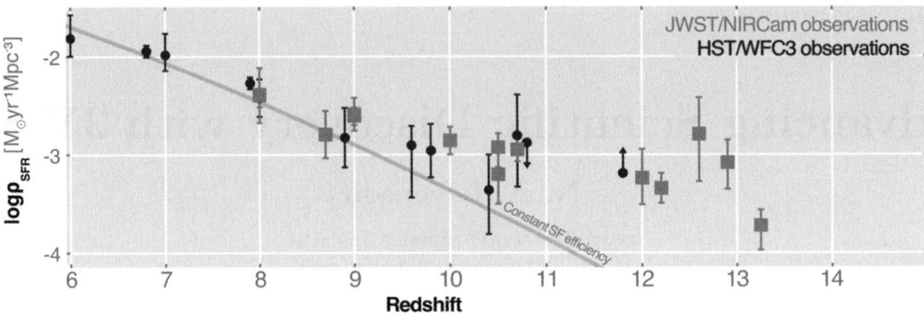

Figure 1. Star formation rate density as a function of redshift. Measurements that make use of JWST/NIRCam are plotted as large squares; measurements from the Hubble Space Telescope's Wide Field Camera 3 (HST/WFC3) are plotted as small circles. Multiple observations beyond $z = 10$ provide signficant measurements. A line of constant star-formation efficiency is plotted in gray. Although it reproduces the observations well at $z < 9$, the high-redshift sources exceed this rate, requiring some more efficient or stochastic mode of star formation. Figure credit: Adamo et al. (2024), after Harikane et al. (2024).

that stellar initial mass functions unlike those of the Milky Way and local galaxies and stochastic star formation may be relevant (Figure 1).

Candidate distant galaxies are identified in deep imaging observations, or directly using slitless spectroscopy capabilities (*e.g.*, Finkelstein et al. (2023); Windhorst et al. (2023); Donnan et al. (2024); Bezanson et al. (2024); Roberts-Borsani et al. (2022); Matthee et al. (2023)) High-redshift galaxies are only detectable at the longest wavelengths. Multi-band photometry yields redshift estimates, but spectroscopy is most reliable, accounting for unusual spectral features or nearby interlopers. Currently, the highest confirmed redshift is $z = 14.32$ (Carniani et al. 2024), from the time only 300 million years after the Big Bang. It is spatially resolved, but does not have any measurable emission lines.

Spectra of galaxy samples frequently show strong emission lines, even at high redshift (Bunker et al. 2024; Meyer et al. 2024). Many abundances measurements are lower than solar (Nakajima et al. 2023; Sanders et al. 2024). Some unusual ratios are detected, notably enhanced nitrogen (Cameron et al. 2023), which may be related to the physical conditions (such as high density) and stellar properties. Line emission also indicates the presence of supermassive black holes in galaxy centers. Given the short timescales available, stellar-mass remnants as seeds together with accretion limited to the Eddington rate are insufficient to produce these massive black holes (Maiolino et al. 2024), assuming that local scaling relations are appropriate to infer black hole mass. Heavy seeds could be primordial—the result of density fluctuations—or the result of direct collapse of supermassive stars. Super-Eddington accretion could proceed episodically and need not be sustained uniformly over time.

JWST is enabling measurements of the era of cosmic reionization, which made the Universe transparent. While the end of cosmic reionization by $z \sim 5$–6 has been established, the onset has been less clear. Direct detection of light from the early epochs demonstrates that the Universe was no longer fully opaque a few hundred million years after the Big Bang. Reionization did not proceed uniformly, and good spatial coverage at high redshifts enables mapping some of the three-dimensional detail of distinct ionized bubbles around clustered galaxies (Chen et al. (2024); Figure 2). A standing problem of reionization had been providing sufficient ionizing photons that escape galaxies. This problem may now be solved, given the large numbers of early, UV-bright galaxies and active galactic nuclei that are evident (Madau et al. 2024).

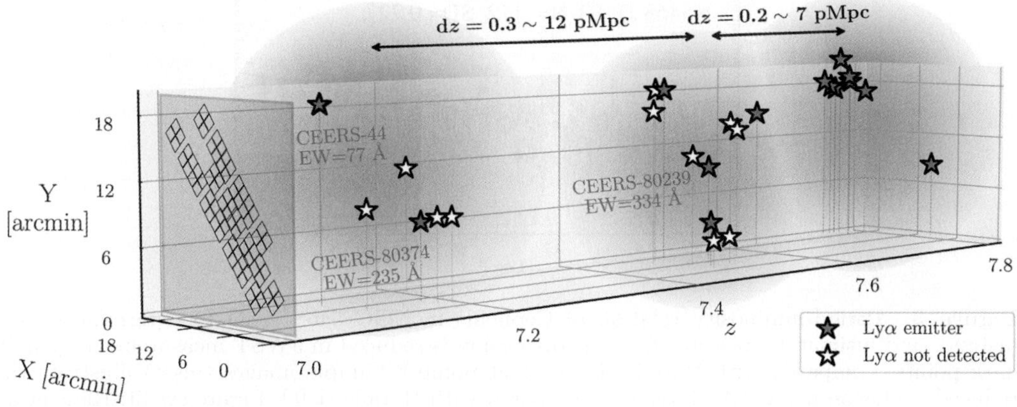

Figure 2. Three-dimensional model of reoinzation (shaded regions) from $z = 7.0$ to 7.8 utilizes wide coverage in the plane of the sky and multiple lines of sight to distant galaxies. The galaxies (star symbols) are clustered, and only some are observed to emit Lyman α (filled symbols). Figure credit: Chen et al. (2024).

3. Galaxies Through Time

Galaxies are signposts for measuring the expansion of the Universe at late times. These expansion rates, expressed as the Hubble constant, H_0, are significantly different from those based on observations of the cosmic microwave background, producing a so-called "Hubble Tension" in some cases. JWST provides access to a Cepheid variable stars over a range of distances, tying purely geometric measurements in the Milky Way (parallax) to Type Ia supernovae, which are observable at even greater distance. JWST reduces confusion compared with previous observations of Cepheids using the Hubble Space Telescope. The dispersion in JWST distance measurements is reduced, and the discrepancy with the early Universe determination persists (Riess et al. 2024). In contrast, Freedman et al. (2024) use JWST observations in multiple techniques of local stellar-based distance measurement—including Cepheids—and find consistency with the Λ-cold dark matter models of the early Universe (Figure 3).

JWST helps to illustrate the physics of star formation in nearby galaxies. The structures out of which stars form are evident, especially in the arms of spiral galaxies. Characteristics include filaments and bubbles, which polycyclic aromatic hydrocarbons (PAHs) in the cold interstellar medium trace (Sandstrom et al. 2023). The new JWST multi-band infrared imaging of the galaxies in this survey (Lee et al. 2023) complement existing multi-wavelength observations to provide a more complete account of the variety of phases of the interstellar medium and their relationshiip to star formation.

JWST observations of resolved stellar populations reveal the history of star formation and enrichment in nearby galaxies. Deep near-infrared observations of the Local Group galaxy Wolf-Lundmark-Melotte (WLM) provide a color-magnitude diagram of nearly 600,000 stars (McQuinn et al. 2024). The derived star formation history indicates a pause after an initial burst of star formation. This pause may be related to the additional heating available after the epoch of reionization, preventing gas cooling for star formation.

4. The Stellar Lifecycle

Astronomers are using JWST to measure all stages of the stellar lifecycle. The early stages—of protostars and disks—are some of the most revealing, especially utilizing high-sensitivity and high-resolution spectroscopy. Observations of background stars probe the envelope of a Class 0 protostar in the Chameleon I star-forming region

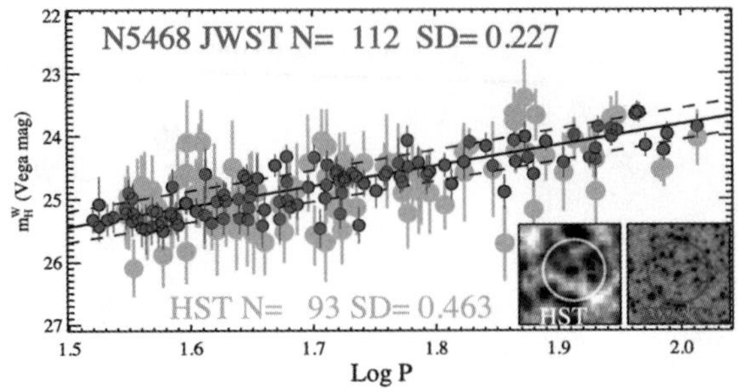

Figure 3. Period-luminosity relation of Cepheids in NGC 5468, used to determine galaxy distance and ultimately measure H_0. The dispersion is reduced in JWST measurements (small, dark points) compared with Hubble (large, light points). Sample images (inset) illustrate the reduced confusion using JWST (right), compared with Hubble (left). Figure credit: Riess et al. (2024).

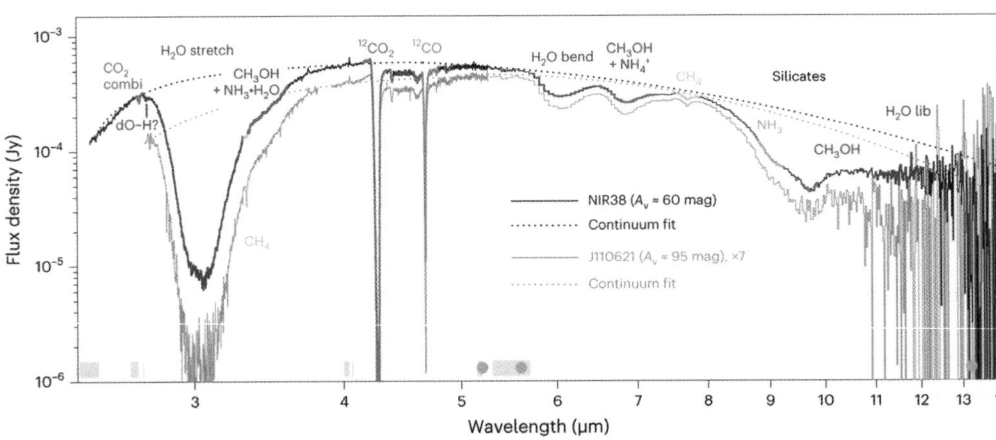

Figure 4. Ices detected in absorption in a dense molecular cloud. Dark and light spectra correspond to two distinct lines of sight toward a protostar. Spectroscopy obtained from NIRSpec and MIRI provides broad wavelength coverage. Figure credit: McClure et al. (2023).

(McClure et al. (2023); Figure 4). Principal ices are measured, with estimates of around 40% of CO frozen in ices. The authors conclude that many species are effectively formed early, including CO_2, NH_3, and CH_4. Other new studies of complex molecules and ions in young protostars show ice and gas features and signatures of outfows (*e.g.*, Yang et al. (2022); Ray et al. (2023); Rocha et al. (2024)).

Debris disks are dynamic (Rebollido et al. 2024; Lawson et al. 2023) and chemically rich (Gáspár et al. 2023). Arabhavi et al. (2024) show the abundant hydrocarbons in the disk around the very low-mass ($0.11 M_\odot$) star, ISO-ChaI 147. Such disks are also interesting as the likley formation sites of terrestrial planets. This example has a relatively high carbon to oxygen ratio. The gas-phase carbon could be enriched with the destruction of carbon-rich (solid) dust grains. Emission is not detected from silicate dust or polycyclic aromatic hydrocarbons.

JWST is also used effectively to examine the end stages of stars. Supernova 1987A (SN1987A) has been well-studied over the decades since its discovery. The characteristics of the progenitor, a blue supergiant, along with the detection of neutrino emission, suggest a neutron star or black hole as a compact remnant. JWST observations show strong evidence for the former. Fransson et al. (2024) measure multiple ionized emission lines in spatially resolved images through the near- and mid-infrared bands, notably forbidden transistions of argon and sulfur. The line ratios do not definitively distinguish the origin as either shocks or photoionization. However, both would result from a neutron star; in the former case, the consequences of a pulsar wind nebula, and in the latter case, either directly from a cooling neutron star or indirectly from a pulsar wind nebula.

5. Other Worlds

JWST provides access to planets within and beyond the Solar System. Although not part of the original motivating science case (with no confirmed examples at the time), exoplanets account for a significant fraction of JWST's observations. JWST's infrared sensitivity is useful for both direct imaging (using coronagraphy) and spectroscopy. The coldest exoplanet imaged so far, Eps Ind Ab, has a temperature around 275K (Matthews et al. 2024). The host star is roughly solar-aged, in contrast to most directly imaged planets that are in younger systems (and therefore warmer and more luminous).

Most of the exoplanet systems studied are very different from the Solar System. M dwarf host stars provide greater contrast with their planets. Hot Jupiters have more substantial signatures, and their shorter orbital periods facilitate observations of multiple transits. Transmission spectroscopy reveals details of the atmospheres of hot, luminous gas giants. Depending on their composition, the tenuous atmospheres preferentially absorb the host starlight during transit. The stability of JWST is helpful to make the difficult differential measurements, at the level of hundreds of parts per million in many cases. The first detection of CO_2 was evident in some of the earliest science observations (JWST Transiting Exoplanet Community Early Release Science Team 2023). The source, WASP-39b is approximately the mass of Saturn although larger, with a temperature around 1100K. CO_2 absorption is strong in the 4.1–4.5μm bandpass. Combined with the absence of methane features, these authors conclude that the atmosphere is metal-enriched. Water is also indicated, at wavelengths $\lambda < 3.6\mu$m.

More detail can be extracted considering the diurnal variation of the atmosphere, rather than assuming that it is uniform. Espinoza et al. (2024) measure differences in the morning and evening atmospheric spectra of WASP-39b. They attribute the larger evening transit depths to higher temperature, a difference of 177K. Circulation in the atmosphere can reproduce these observations, specifically with a cloudy morning atmosphere and a clear evening (Figure 5).

Rocky planets exhibit weaker atmospheric signatures, if they exist at all. Hu et al. (2024) find evidence for a volatile atmosphere in the terrestrial planet 55 Cancri e, based on on near- and mid-infrared observations during the secondary eclipse. The spectral resolution at shorter wavelengths usefully indicates the presence of volatiles, such as CO or CO_2. An alternative is that vaporized rock of a lava world provides a very thin atmosphere. Modeling the rock composition with bulk silicates similar to Earth cannot fit the data as well. The overall temperature is too high because the molten lava does not adequately redistribute heat, and absorption features of the volatiles are absent.

In other terrestrial exoplanets, observations are inconclusive. In addition to the weakness of atmospheric themselves, other effects can mimic atmospheric signatures. The super-Earth GJ 486b, for example, does exhibit features distinct from a flat line, and those are suggestive of water (Moran et al. 2023). However, either the atmosphere or star spots in the host could be responsible for the features. Observations at shorter wavelengths could distinguish between these two possibilities.

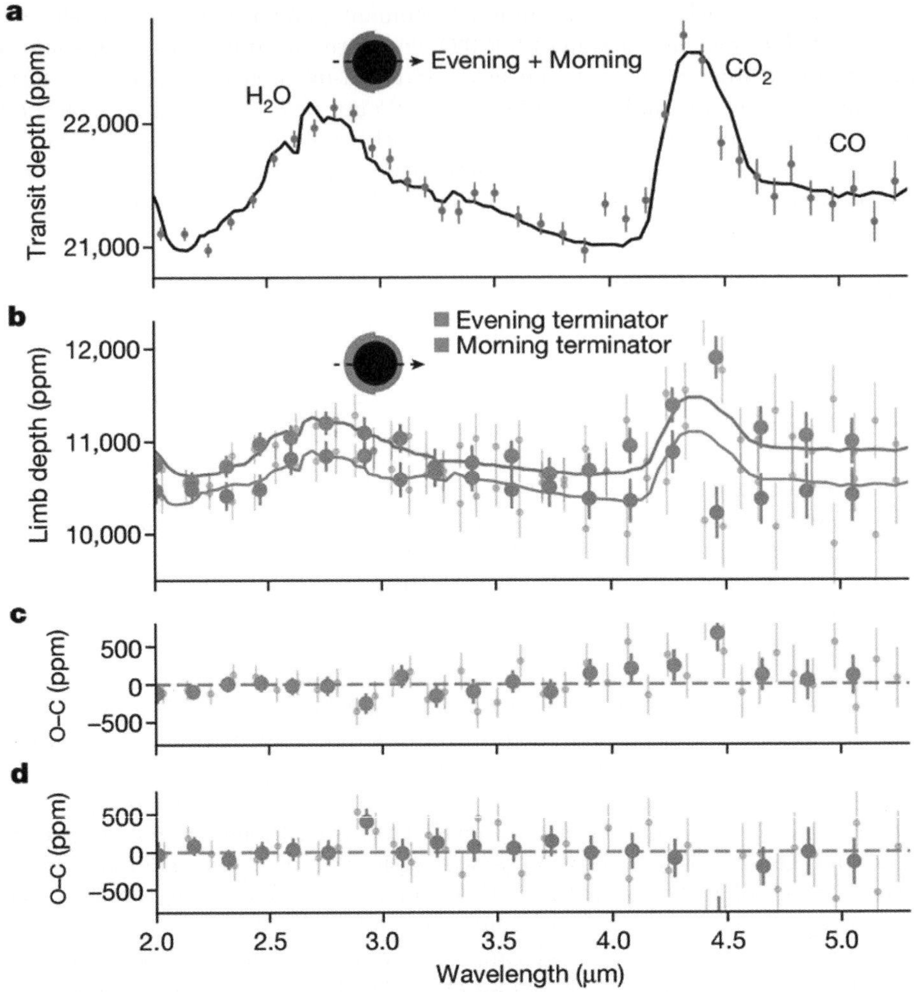

Figure 5. Transmission spectra of WASP-39b. The top panel (a) shows the total transit depth. Panel (b) distinguishes the spectra obtained at the evening terminator (orange or darker, upper points) from the morning terminator (blue or lighter, lower points). Large points are binned to lower resolution. Curves show best-fitting models in each case. Residuals of these models are shown in panels (c; evening) and (d; morning). Figure credit: Espinoza et al. (2024).

6. Open and Inclusive Access to JWST

JWST remains open to a global science community. A competitive peer review process recommends time for new observations. The last completed annual round, Cycle 3, included more than 6,000 unique investigators across 57 countries.

A dual anonymous process does not reveal the reviewing panel to the proposers, and the evaluation occurs without knowledge of the proposing teams. This process concentrates the assessment on the science, not the scientists. We now find equal proposal success by gender, and roughly 10% of principal investigators (PIs) are students. Nearly 50% of PIs are new in this role to Space Telescope Science Institute, which also conducts the Hubble Space Telescope (HST) allocation process. Relative to previous HST review cycles in which proposers were identified, the distribution of PIs is shifted to earlier career stages. An overview of the telescope allocation results is a regular part of JWST Users Committee

(JSTUC) reports, which are publicly available through the meeting agenda pages: `https://www.stsci.edu/jwst/science-planning/user-committees/jwst-users-committee`.

All data become fully open for community use, with a maximum exclusive access period of 12 months for certain types of observations. Data are available through the Barbara A. Mikulski Archive for Space Telescopes (MAST). A variety of tools support queries that may be general or JWST-specific, described at `https://archive.stsci.edu/missions-and-data/jwst`. STScI has taken steps to make substantial data available early during the JWST mission lifetime. The Director's Discretionary Early Release Science (ERS) effort established 13 programs with zero exclusive access. These programs covered a wide range of instrument observing modes and science cases. They were preferrentially observed in the initial science phase, to provide immediate public access to JWST data. The astronomical community has responded quickly, producing over 500 refereed publications already in the literature just two years after the release of first science data.

STScI provides extensive resources to support research with JWST. A standard pipeline delivers reduced data that is adequate in many cases. The pipeline code is open source, and users may run the pipeline with different parameters or modify the code itself. Specific tools support analysis of processed JWST data. Information about these tools and various modes of training (such as video resources and Python notebooks) are available through `https://www.stsci.edu/jwst`.

7. Future Opportunities

7.1. *General Approaches*

The discoveries with JWST open new areas for exploration. The first two years of data reveal exceptional individual examples of various types of astrophysical sources. Often, however, systematic observations that would show the full range of the class of objects or support general conclusions are lacking. To account quantitatively for the formation and evolution of the first stars and galaxies, for example, additional observations are necessary. One path is to build robust samples of high-redshift galaxies. These samples can be subdivided by redshift, mass, or other property. Beyond increasing the number of sample galaxies, spectroscopy best provides physical diagnostics, despite the more limited feasible of measuring large numbers of sources. Existing deep imaging surveys provide excellent candidates for spectroscopy.

7.2. *Strategic Director's Discretionary Programs*

STScI convened two international community working groups in 2023 to provide input on strategic programs that may use director's discretionary time (DDT) allocations. These focused on: long-term variability and monitoring; and exoplanets. Such strategic DDT programs open fundamental scientific opportunity in ways that would be unlikely to emerge from the standard time allocation process. They enable multiple investigations from diverse teams using high-quality, science-ready data sets. Extending beyond the perspectives of the working group members, these committees solicited broad community input through surveys, white papers, and town halls. The charge of the former was broad, considering operational and policy changes to enable research across relevant areas. A specific observational recommendation (Jha et al. 2024) would utilize high-redshift transients to probe the early Universe. A Population III supernova would likely be the most observable evidence of the first generation of stars. Such stars are likely very massive, formed in metal-free environments, with distinct explosion mechanisms (as "pair-instability supernovae") and observable properties. An observing strategy could take advantage of the continuous viewing zone for optimal cadence, or instead concentrate on galaxy clusters, to use gravitaional lensing to reach the highest redshifts.

STScI is implementing the primary recommendation of exoplanet working group (Redfield et al. 2024), a large-scale survey of rocky M-dwarf exoplanets. Do such planets have atmospheres? What conditions make them likely? What variety do they exhibit? Systematic data from a well-selected sample will begin to answer these questions. The observational program will measure thermal emission from more than a dozen systems, taking advantage of JWST's unique infrared strengths to probe planets around these low-mass stars. The program will specifically rely on measurements at 15μm in secondary eclipse. HST will provide ultraviolet observations of the host stars, as stellar activity also affects the persistence and properties of planetary atmospheres. Community perspectives on the program—including advice on target selection, data needs, and equitable community interactions—continue with a Science Advisory Council.

7.3. *Beyond JWST*

Flagship missions like JWST require years of development. Many of their greatest discoveries were not foreseen before they existed. Nonetheless, current scientific questions drive the demand for future capabilities. Despite the great progress we anticipate in the investigation of exoplanets with JWST, its contrast and wavelength coverage limit its ability to detect evidence of life beyond the Solar System. A large, stable space telescope, sensitive across the ultraviolet–near-infrared regime is required. The leading recommendation for a space facility from the US Decadal Survey on Astronomy and Astrophysics (National Academies 2023), now known as Habitable Worlds Observatory (HWO), would respond to these open questions. NASA has is supporting the early development of HWO, crucially including maturation of the essential technology.

The ambitions for HWO are civilization-scale and entail global participation. The basic capabilities open broad opportunities across all fields of astrophysics and planetary science. Resolution at a minimum of 200 pc scales at all redshifts, for example, will be possible, enabling studies of galaxy evolution. Spatially resolved emission and absorption spectroscopy of the intergalactic medium will map the baryon cycle. Accurate photometry and astrometry will enable analysis of stellar populations and physics to signficant distances. Development of these more detailed, specific science cases will set mission requirements. in the meantime, JWST offers opportunity for discovery across the full breadth of astrophysics.

References

Adamo, A., Atek, H., Bagley, M. B., et al. 2024, arXiv:2405.21054
Arabhavi, A. M., Kamp, I., Henning, T., et al. 2024, *S*cience, 384, 1086
Bezanson, R., Labbe, I., Whitaker, K. E., et al. 2024, *ApJ*, 974, 92
Böker, T., Beck, T. L., Birkmann, S. M., et al. 2023, *PASP*, 135, 038001
Bunker, A. J., Cameron, A. J., Curtis-Lake, E., et al. 2024, *A&A*, 690, A288
Cameron, A. J., Katz, H., Rey, M. P., et al. 2023, *MNRAS*, 523, 3516
Carniani, S., Hainline, K., D'Eugenio, F., et al. 2024, *Nature*, 633, 318
Chen, Z., Stark, D. P., Mason, C., et al. 2024, *MNRAS*, 528, 7052
Donnan, C. T., McLure, R. J., Dunlop, J. S., et al. 2024, *MNRAS*, 533, 3222
Doyon, R., Willott, C. J., Hutchings, J. B., et al. 2023, *PASP*, 135, 098001
Espinoza, N., Steinrueck, M. E., Kirk, J., et al. 2024, *Nature*, 632, 1017
Finkelstein, S. L., Bagley, M. B., Ferguson, H. C., et al. 2023, *ApJL*, 946, L13
Fransson, C., Barlow, M. J., Kavanagh, P. J., et al. 2024, *Science*, 383, 898
Freedman, W. L., Madore, B. F., Jang, I. S., et al. 2024, arXiv:2408.06153
Gardner, J. P., Mather, J. C., Abbott, R., et al. 2023, *PASP*, 135, 068001

Gáspár, A., Wolff, S. G., Rieke, G. H., et al. 2023, *Nature Astronomy*, 7, 790
Harikane, Y., Nakajima, K., Ouchi, M., et al. 2024, *ApJ*, 960, 56
Hu, R., Bello-Arufe, A., Zhang, M. et al. 2024, *Nature*, 630, 609
Jha, S., Casetti-Dinescu, D. I., Bernsein, G. M., et al. 2024, arXiv:2405.12297
JWST Transiting Exoplanet Community Early Release Science Team 2023, *Nature*, 614, 649
Lawson, K., Schlieder, J. E., Leisenring, J. M., et al. 2023, *AJ*, 166, 150
Lee, J. C., Sandstrom, K. M., Leroy, A. K., et al. 2023, *ApJL*, 944, L17
Madau, P., Giallongo, E., Grazian, A., et al. 2024, *ApJ*, 971, 75
Maiolino, R., Scholtz, J., Witstok, J., et al. 2024, *Nature*, 627, 59
Matthee, J., Mackenzie, R., Simcoe, R. A., et al. 2023, *ApJ*, 950, 67
Matthews, E. C., Carter, A. L., Pathak, P., et al. 2024, *Nature*, 633, 789
Meyer, R. A., Oesch, P. A., Giovinazzo, E., et al. 2024, *MNRAS*
McClure, M. K., Rocha, W. R. M., Pontoppidan, K. M., et al. 2023, *Nature*, 7, 431
McQuinn, K. B. W., B. Newman, M. J., Savino, A., et al. 2024, *ApJ*, 961, 16
Moran, S. E., Stevenson, K. B., Sing, D. K., et al. 2023, *ApJL*, 948, L11
Nakajima, K., Ouchi, M., Isobe, Y., et al. 2023, *ApJS*, 269, 33
National Academies of Sciences, Engineering, and Medicine 2023, Pathways to Discovery in Astronomy and Astrophysics for the 2020s. Washington, DC: The National Academies Press
Ray, T. P., McCaughrean, M. J., Caratti o Garatti, A., et al. 2023, *Nature*, 622, 48
Rebollido, I., Stark, C. C., Kammerer, J., et al. 2024, *AJ*, 167, 69
Redfield, S., Batalha, N., Benneke, B., et al. 2024, arXiv:2404.02932
Rieke, M. J., Kelly, D. M., Misselt, K., et al. 2023, *PASP*, 135, 028001
Riess, A. G., Anand, G. S., Yuan, W., et al. 2024, *ApJL*, 962, L17
Rigby, J., Perrin, M., McElwain, M., et al. 2023, *PASP*, 135, 048001
Roberts-Borsani, G., Morishita, T., Treu, T., et al. 2022, *ApJL*, 938, L13
Rocha, W. R. M., van Dishoeck, E. F., Ressler, M. E., et al. 2024, *A&A*, 683, A124
Sanders, R. L., Shapley, A. E., Topping, M. W., et al. 2024, *ApJ*, 962, 24
Sandstrom, K. M., Koch, E. W., Leroy, A. K., et al. 2023, *ApJL*, 944, L8
Windhorst, R. A., Cohen, S. H., Jansen, R. A., et al. 2023, *AJ*, 165, 13
Wright, G. S., Rieke, G. H., Glasse, A., et al. 2023, *PASP*, 135, 048003
Yang, Y.-L., Green, J. D., Pontoppidan, K. M., et al. 2022, *ApJL*, 941, L13

The First Chapters of Our Cosmic History with JWST
Proceedings IAU Symposium No. 391, 2025
T. Nanayakkara & M. Maseda, eds.
doi:10.1017/S1743921325000067

Fast Forming and Fast Quenching: Massive Quiescent Galaxies at $z > 3$

Karl Glazebrook

Centre for Astrophysics and Supercomputing, Swinburne University of Technology,
PO Box 218, Hawthorn, VIC 3122, Australia
email: kglazebrook@swin.edu.au

Abstract. Massive quiescent galaxies (MQGs) at redshifts $z > 3$ present a profound challenge to our understanding of galaxy formation and evolution. These galaxies undergo rapid and intense star formation followed by swift quenching very early in cosmic history, resulting in the formation of compact galaxies with half-light radii smaller than 1 kpc. New observations, particularly from the James Webb Space Telescope (JWST), have rapidly advanced our knowledge of their abundance, possible formation pathways, and physical properties. However, current theoretical models still struggle to replicate the observed number densities and characteristics of these galaxies. This review attempts to consolidate recent observational breakthroughs, discusses the comparison with theoretical models, and highlights key unresolved questions. An especially interesting development is the discovery of sub-populations of even older and early forming (at $z > 7$) quiescent galaxies which pose particular challenges and add to the ledger of the 'too much too soon' galaxy and AGN problems revealed by JWST. I outline my own views on key future research directions to unravel the mysteries of MQGs and their role in the broader context of cosmic evolution.

Keywords. galaxies, evolution, quenching

1. Introduction

Massive quiescent galaxies (MQGs) are a cornerstone for understanding galaxy formation and evolution. Their presence at high redshifts (with the current focus now lying at $z > 3$) challenges models of galaxy evolution within the framework of ΛCDM cosmology. Specifically, the formation of MQGs in the early universe necessitates the rapid assembly of stellar mass within short cosmic timescales, often less than a billion years, which has to be consistent with the hierarchical growth of dark matter halos predicted by ΛCDM.

Early studies confirmed spectroscopically the presence of massive galaxies at $z > 1.5$ in abundances greater than the contemporaneous predictions of hierarchical models (Glazebrook et al. 2004). Many of these were found to be quiescent (with ages of up to several Gyr) and extremely compact (McCarthy et al. 2004; Daddi et al. 2005; Trujillo et al. 2006; van Dokkum et al. 2008; Damjanov et al. 2009) with effective radii 1–2 kpc; a population at $z \sim 2$ that became referred to as 'red nuggets'. Quiescent galaxies present the most extreme test for galaxy formation in massive halos as they are the first to form, and then quench, and *then* evolve their stellar populations; their age effectively pushes the abundance constraints to higher redshifts than that at which they are observed.

The evolving density of dark matter halos at high redshifts limits the availability of massive halos capable of hosting MQGs. Theoretical models, such as those presented by Dekel et al. (2009) and Zolotov et al. (2015), proposed mechanisms like gas compaction and AGN feedback to explain the rapid quenching of star formation. High observed abundances would imply either extremely efficient baryonic processes or gaps in our understanding of halo formation and galaxy-halo interactions. These challenges make MQGs a critical testing ground for galaxy formation and cosmological models and motivated the extensions to higher redshifts ($z > 3$) where the abundance of massive objects is an even more existential test. This review aims to summarise the current state of the field, highlighting key observations, theoretical challenges, and future directions.

2. Massive Quiescent Galaxies: Definitions and Key Properties

MQGs at $z > 3$ are characterized by their lack of significant star formation activity for over 100 Myr and stellar masses $\gtrsim 10^{10} M_{\odot}$. They are commonly identified using rest-frame UVJ color diagrams (Straatman et al. 2014) (or directly from star formation rate (SFR) measures), photometric redshifts, and spectral energy distributions (SEDs). Compact morphologies and high stellar densities are hallmarks of these systems. Studies reveal that MQGs exhibit smaller effective radii compared to star-forming galaxies of similar mass (Straatman et al. 2015), suggesting that they form through dissipative processes.

2.1. *Pre-JWST Observations*

The first discoveries of massive quiescent galaxies at $z \sim 2$ were made using deep ground-based near-infrared imaging and Hubble Space Telescope (HST) data. These galaxies were identified as high-redshift quiescent systems from colours and spectra (Cimatti et al. 2004; McCarthy et al. 2004; Brammer et al. 2009; Williams et al. 2010). Subsequent studies confirmed their early-type morphology, small effective radii ($\sim$ 1–2 kpc) and high stellar densities (Daddi et al. 2005; Trujillo et al. 2006; Kriek et al. 2008; Cimatti et al. 2008; van Dokkum et al. 2008; Damjanov et al. 2009).

Theoretical models to explain these around 2014–2015 predicted that the formation channels of such systems were through gas-rich mergers and dissipative compaction processes (Dekel et al. 2009; Zolotov et al. 2015; Wellons et al. 2015). These models suggested that intense star formation during compaction events leads to the rapid consumption of gas and subsequent quenching, resulting in compact quiescent systems. While these models successfully explained some of the observed properties of $z \sim 2$ MQGs, they struggled to account for their high number densities (Wellons et al. 2015). This number density is predicted to evolve extremely fast, for example model stellar mass functions (Genel et al. 2014) predict a factor of 50-fold drop in the space density of $\sim 10^{11}$ $M_{\odot}$ stellar mass galaxies between $z = 2$ and $z = 4$. This reflects the evolution of the underlying halo abundance (at halo mass $\sim 10^{13}$ $M_{\odot}$) in ΛCDM cosmologies. Further to this the frequency of quiescent galaxies among the massive populations also drops and none were predicted to exist at all at $z \geq 4$ (Wellons et al. 2015).

This motivated observational extensions to $z > 3$; for example Straatman et al. (2014) published a sample of photometric candidates (with a notably high number density) at $z \sim 4$. These were again UVJ-selected and exhibited even more compact sizes than at $z \sim 2$, and also showed early-type morphologies. The availability of new infrared spectrographs on Keck and VLT then led to spectroscopic confirmations (Glazebrook et al. 2017; Schreiber et al. 2018a; Valentino et al. 2020; Forrest et al. 2020; Forrest et al. 2020) — at least of the brighter objects. The typical spectrum of such MQGs turned out to be 'post-starburst'; (PSB) i.e. ages of several hundred Myr with spectra dominated by Balmer absorption lines (H$\delta\gamma\beta$) and the Balmer break. It is important to note that these are very different spectra to the majority of low-redshift MQGs ($z < 1$) which are instead many Gyr old and dominated by lines such Ca H & K, the G-band, Mg-b and the 4000Å break. The fractions of 'Hδ strong' MQGs at $z = 0$ for example is only a few

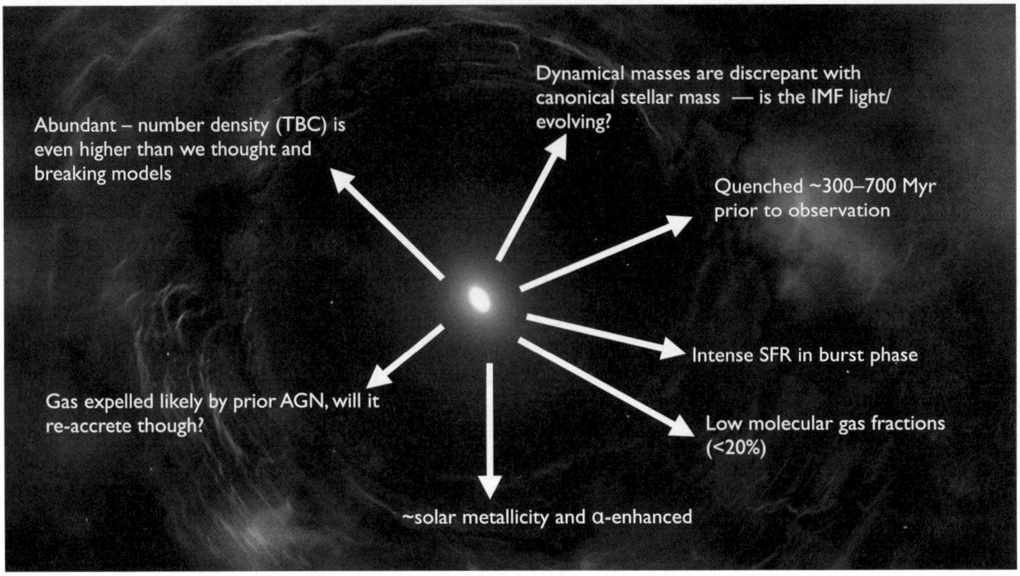

Figure 1. Key currently known properties of $z > 3$ massive quiescent galaxies with an artist's impression of their appearance (Credit: J. Josephides, Swinburne)

percent, rising to 50% at $z = 1$ (Le Borgne et al. 2006) and it is now 100% at $z \sim 4$. This reflects of course that the Universe is younger at higher redshift and so there is less time to form galaxies, quench galaxies and developed Gyr-age features.

High number densities observed remained a clear problem with only a subset of updated galaxy formation models predicting close to the right abundances (Schreiber et al. 2018a; Merlin et al. 2019; Valentino et al. 2020). A simple theoretical argument illustrates where this problem arises. One can infer the halo mass from the observed number density (effectively assuming a rank-ordered halo model), and then by multiplying this by Ω_b/Ω_m one can compute the halo baryonic mass; the ratio to the stellar mass gives the 'galaxy formation efficiency' (Glazebrook et al. 2017). These values typically come out around $\sim 30\%$ for MQGs at $z > 3$ which is very similar to massive galaxies at $z \sim 0$ (Baldry et al. 2008) whereas one would intuitively expect significant evolution as galaxies are less built up. An important caveat was the incompleteness of ground-based spectroscopic samples rendering this comparison highly uncertain. The deepest ground-based spectroscopic campaigns were only confirming $< 50\%$ of galaxies. For example Schreiber et al. (2018a) confirmed 12 out of 24, two of which were interlopers. Even some quite bright sources ($K < 22.5$) eluded spectroscopic confirmations; possibly because they had no strong spectral features in the limited ground H and K-band spectral windows (this proved true).

3. Observational Progress in High-Redshift MQGs with JWST

The advent of the capabilities of JWST (Rigby and et al. 2023), with it spectacular near-infrared imaging (NIRCam; Rieke et al. (2023) and spectroscopic (NIRSpec; Böker et al. (2023)) performance is rapidly revolutionising the topic; especially at the highest redshifts. NIRCam can be used to search for MQGs and also to characterise the morphology, with greater resolution, signal/noise and wavelength coverage than HST, of known ones. This has greatly expanded our knowledge of the key properties of MQGs as summarised in Figure 1.

3.1. *Number Density*

The number densities of MQGs at $z > 3$ remain a critical observational constraint with JWST showing that previous ground-based selections were incomplete. A NIRCam photometric search by Carnall et al. (2023a) found a surprising number at $3 < z < 5$ in only a 30 arcmin^2 area; a factor of 3–5 $\times$ higher than previous ground-based searches. This reflected a combination of improved deep imaging data at $> 2\mu$m giving better detection and photometric redshifts and the ability of JWST to resolve objects overlapping with foreground galaxies in the sky. Alberts et al. (2024) extended the wavelength coverage of searches to include deep MIRI imaging though they noted that no large bias was induced by pure NIRcam selection. They found one candidate at an astonishing photometric redshift of 5.33. It should be noted that with small areas surveyed cosmic variance could be a huge problem; MQGs likely lie in extremely clustered environments. However JWST surveys are rapidly getting larger which should mitigate this. These much higher number densities have exacerbated the problems of theoretical models; taking the extreme case at $z = 5$ the observed number densities are 50–500$\times$ too high (Weller et al. 2024)! Spectroscopic confirmation of redshifts has so far focussed on ground-based selected samples. Nanayakkara et al. (2024) completed the sample of Schreiber et al. (2018a) with a 100% redshift confirmation rate in a Cycle 1 program; they concluded that the purity of the sample was 90% but also noted that many objects showed emission lines which they attributed to AGN and not star-formation. Many of these emission lines were in regions outside the ground-based spectroscopic windows.

A related question is the abundance of potential progenitor populations (Valentino et al. 2020), while dusty starbursting sub-mm galaxies at $z > 6$ exist (e.g. (Riechers et al. 2013)) and are excellent candidates their space density has been hard to estimate. When you select on extreme star formation you have to consider duty cycle issues. Interestingly, recently the JWST FRESCO grism survey has found through rest-frame optical emission line selection an additional population of 'massive optically dark star forming galaxies' (Xiao et al. 2024). This new population have stellar masses $> 10^{11}$ $M_\odot$ which are higher than other candidate progenitors and in themselves present their own 'high galaxy formation efficiency' problem.

3.2. *Spectroscopic analysis and Stellar Ages*

Spectroscopic analyses of $z > 3$ MQGs reveal post-starburst characteristics, with typical stellar ages of 300–700 Myr (Schreiber et al. 2018a; Forrest et al. 2020; Carnall et al. 2023a; Nanayakkara et al. 2024). In many cases the JWST spectra have revealed exquisite PSB spectra (Carnall et al. 2023b), as good as spectra of nearby galaxies. These results suggest that star formation quenched rapidly, following intense starbursts. The age constraints imply formation redshifts $z \gtrsim 5$, consistent with observations of dusty star-forming progenitors (Straatman et al. 2014; Valentino et al. 2020) and consistent with the predictions of the current generation of galaxy formation models for similarly selected objects (Nanayakkara et al. 2024).

However there are exceptions. Glazebrook et al. (2024) reported a 10^{11} $M_\odot$ stellar mass $z = 3.2$ MQG ('ZF-UDS-7329') that was 1.5 Gyr old; so whose stellar populations formed at $z \sim 11$. Its age was quite distinct from the rest of the sample of Nanayakkara et al. (2024). This age was confirmed by Carnall et al. (2024) who obtained a beautiful high-resolution spectrum of the object with NIRSpec showing clear Gyr-age spectroscopic features. MQG PSB objects have also now been reported at $z > 4$; despite their younger ages this also implies formation at $z > 7$ and hence high formation efficiencies (Carnall et al. 2024; Graaff et al. 2024). Thus there appears to be a distinct sub-population of 'early formers'; for example this is apparent to my eye in Figure 8 of the very nice compilation of Setton et al. (2024).

For comparison of the number density with halo models, one can not just assume that the galaxy forms all at once at the formation redshift; instead one must compare with a detailed hydrodynamical simulations that tracks star formation and halo mergers that both build stellar mass. These comparisons suggest that current models can not easily explain galaxies like ZF-UDS-7329 (Glazebrook et al. 2024). Carnall et al. (2024) and Turner et al. (2024) argued that considering the full star formation history of ZF-UDS-7329 which was extended over 200–400 Myr it was just consistent with the maximum allowed halo mass at each prior time step *if the galaxy formation efficiency was 50–100%.* In my view this is correct but is (i) still physically extreme and (ii) not a fully consistent physical model as this does not connect halos across time steps. While this comparison is interesting to study further I think is clear is that the main limitation of all these works is that only a handful of early formers are known which are all found by disparate methods. Wider surveys are urgently needed to firmly establish their abundance.

Spectroscopy also provides metallicity and elemental abundance information; the MQGs appear to be $\sim$ solar metallicity (Glazebrook et al. 2024; Nanayakkara et al. 2024; Carnall et al. 2024) and enhanced in α-elements such as Mg (Carnall et al. 2024); the latter would be consistent with a rapid formation event followed by quenching. The rapidness of the formation could be further constrained this way and would have important implications for model comparisons.

3.3. *Sizes and morphology*

High-resolution imaging from JWST reveals that MQGs at $z > 3$ typically exhibit elliptical morphologies with high bulge:disk ratios. The availability of F444W imaging (significantly redder than HST's reddest F150W band) allows the determination of size and morphology in bands that probe the rest-frame stellar mass. This removes the nagging doubt from HST days that the compact appearance could have been dominated by single small UV clumps. The resolution of JWST is also superior and we now have well measured and tight size-mass diagrams (Ito et al. 2024; Wright et al. 2024). Our group's work analysing the NIRCam images of the Nanayakkara et al. sample is shown in Figure 2 (Kawinwanichakij et al.; in prep.). The typical half-light radius of a 10^{11} $M_\odot$ galaxy is 800 pc! This suggests a highly dissipative formation processes, likely involving gas-rich mergers or compaction (Zolotov et al. 2015). Significant size growth must also happen between $z = 4$ and $z = 2$ where the classical 'red nugget' population is observed. This is likely to occur through two processes: (i) growth of their stellar halos via minor mergers (van Dokkum et al. 2010), and (ii) the addition of *new* MQGs that are born larger. The latter process must happen as the number density at $z = 2$ is several times higher than at $z \sim 4$. The location of the $z = 4$ MQGs today, presumably in the cores of some existing galaxy population, is an interesting question for further work as it has been for $z = 2$ objects (e.g. Trujillo et al. (2014); Hon et al. (2022)).

JWST also permits resolved analysis of spectral energy distributions; for example Setton et al. (2024) presents a MQG at $z = 3.97$ with a redder core than the surrounding envelope, inferred to be older. We can also look at the the correlation of the residuals in the size–mass relation with third parameters. Our analysis of the relation between size and age (Kawinwanichakij et al.; in prep.) suggests that somewhat paradoxically the early formers may be larger and more disky (for example the Glazebrook et al. (2024) object is one of the the largest in that sample). All this provides ample material for more comparisons with the predictions of simulations.

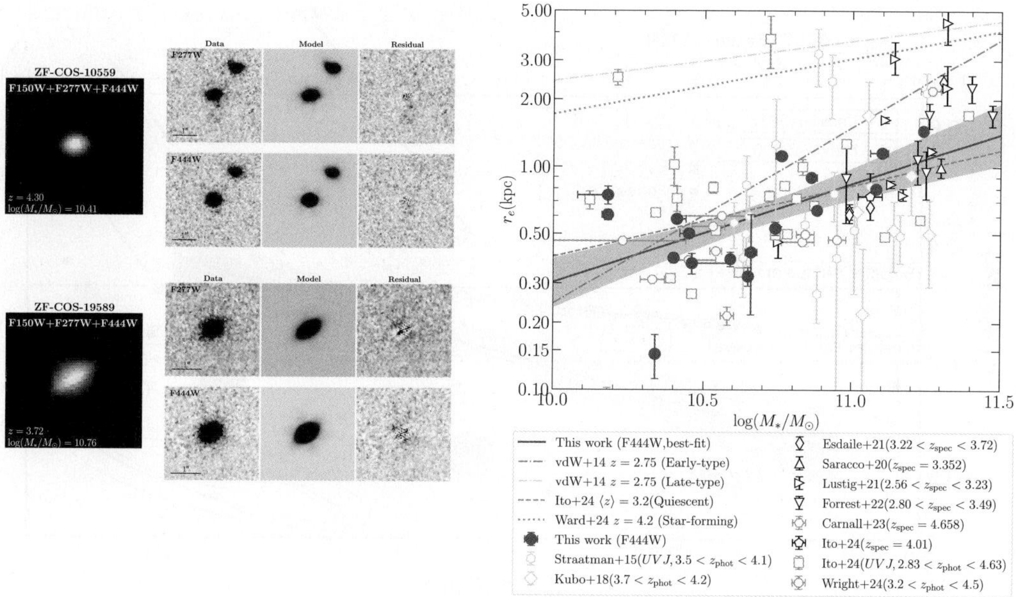

Figure 2. Size–mass relations of $3.0 < z < 4.3$ spectroscopically confirmed massive quiescent galaxies by our group from Kawinwanichakij et al.; in prep. This is based on the sample of Nanayakkara et al. (2024). On the left shows two example objects (NIRCam data) with RGB images and grey scale images showing the data, the model and the residuals after model subtraction. The right panel shows the tight effective radius–mass relation derived (F444W band which is equivalent to rest frame 1μm which is pretty close to stellar mass) and compared with other work at this and lower redshifts.

3.4. *Clustering*

MQGs at high redshift are often found in dense environments, it is quite common to find close companions at very similar redshifts (Schreiber et al. 2018b; Suzuki et al. 2022; Turner et al. 2024) suggesting strong clustering; however quantitive clustering measurements are still quite noisy due to small sample sizes (Magliocchetti et al. 2023). Protocluster candidates harbouring MQGs have been identified (Tanaka et al. 2024). Here, a $z = 4.01$ protocluster in the SXDS field contains five MQGs within 1 physical Mpc, implying a halo mass exceeding $10^{13}\mathrm{M}_\odot$. These observations suggest that environmental quenching plays a significant role in MQG formation. However, Tanaka et al.'s comparison suggested simulations fail to reproduce such clustering properties, with no equivalent structures found in the simulated volume.

Wider area surveys provide a tremendous opportunity for improved clustering measurements both for the auto-correlation of MQGs and their cross-correlation with lower mass and star forming galaxies. This will provide a second route to estimating the host halo masses and hence testing potential tensions, so will be a very important future measurement.

3.5. *Dynamical masses and IMF*

One of the triumphs of the ground-based near-infrared spectroscopy was measurement of the velocity dispersion of some MQGs using the spectroscopic absorption lines (Esdaile et al. 2021; Forrest et al. 2022). They are found to exhibit high velocity dispersions ($\sigma \sim 300 - 400$ km s^{-1}), as was expected from their nature as compact, massive systems. Simple virial estimates allow the determination of dynamical masses which

13

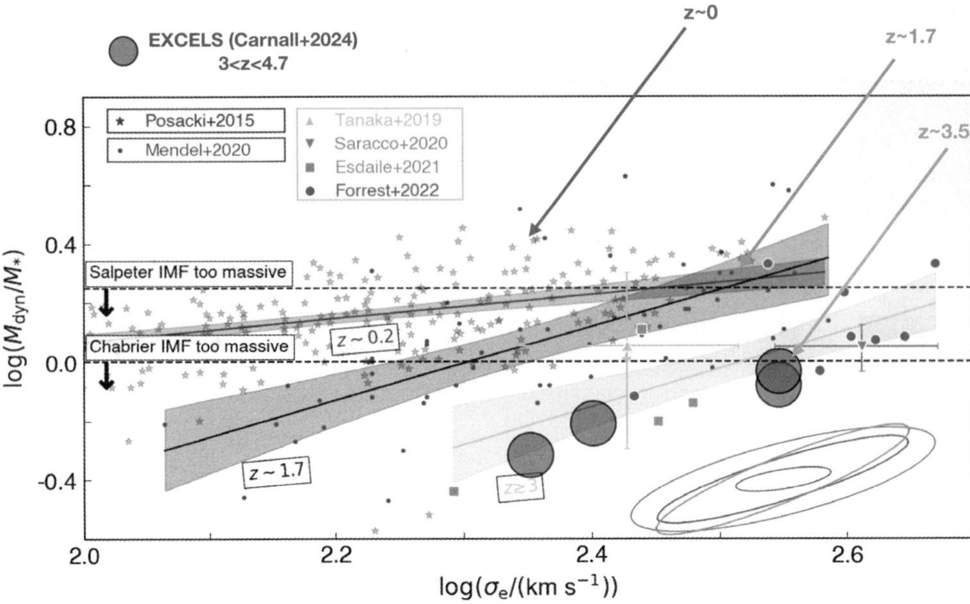

Figure 3. Ratio of dynamical mass to stellar mass adapted from Forrest et al. (2022) where I have added the JWST values I have computed from Table 3 of Carnall et al. (2024) as large red circles. This shows the evolution of this ratio from low-z to $z \sim 4$. The legends show data sources, coloured bands show fits and errors in the relevant range and the ellipses at bottom right (all from the Forrest et al. paper) show the typical errors at $z \sim 3.5$. The key point is this simple exercise show good agreement of JWST with previous ground-based results for this trend.

broadly confirms their stellar masses. This retired one worry that previous conclusions could have been awry due to erroneous stellar masses, for example due to grossly incorrect Initial Mass Funciton (IMF) assumptions. This has now been turned around and these papers have compared dynamical mass to stellar mass to look at IMF evolution. Previously it was found that $z \sim 2$ MQGs had a 'lighter IMF' than at $z \sim 0$ (Mendel et al. 2020). Now it seems that the IMF at $z \sim 4$ may be lighter still and in some cases sub-Chabrier (the Chabrier (2003) IMF being the normal assumption). These are at most factor of two effects on the stellar mass and while important for the understanding of stellar populations may not be large enough to rectify some of the fundamental problems of 'too massive too soon'. I have taken the liberty for this review of converting the size and velocity dispersions from Carnall et al. (2024) in to dynamical masses (Figure). Remarkably these JWST values align well with previous results for the sub-Chabrier trend. The future of dynamical studies is bright with JWST allowing 2D kinematic measurements (d'Eugenio et al. 2024; Pérez-González et al. 2024) improving on 1D velocity dispersions for mass modelling, though this 2D modelling has yet to appear in the literature.

3.6. *Other interesting properties*

A few brief remarks on other interesting properties. First it is widely thought that AGN are likely quenching mechanisms for early MQGs (e.g. Kurinchi-Vendhan et al. (2024)) as there is less time for other mechanisms (e.g. environment) to develop, so it is natural to look for observational AGN signatures. Interestingly a high fraction of $z > 3$ main sequence massive galaxies show emission lines consistent with AGN (Martínez-Marín et al. 2024), the evidence comes from the line widths, the high [OIII]/Hβ ratios (that strongly indicates the ionising source is not star formation) and the line luminosities. One even had X-ray emission though it should be noted that the general lack of X-ray emission is not especially constraining at these redshifts due to the limitations in sensitivity of current telescopes.

An obvious question is whether the fuel for star formation is retained? Again this requires very deep observations, now with sub-mm telescopes such as ALMA, to search for molecular gas. Observations (Suzuki et al. 2022) in the [CI] line show no detections implying MQGs have molecular gas fractions several times lower than main sequence galaxies at the same redshifts with limits of molecular gas mass fractions $\lesssim 20\%$. The sub-mm continuum limits from that paper also confirms the quiescence (i.e. no obscured star formation is detected). The main limitations of this are sample size — only four MQGs were observed requiring up to 4 hours per target. Interestingly JWST spectra reveals an abundance of *neutral* gas as the Na D line absorption is clearly enhanced in nearly all objects (Nanayakkara et al. 2024). At $z \sim 2$ this has also been seen and has been interpreted as an AGN driven outflow signature (Belli et al. 2024). Na D kinematics at $z > 3$ has now been spatially resolved by JWST Integral Field Spectroscopy in two objects (d'Eugenio et al. 2024; Pérez-González et al. 2024); the kinematics are quite distinct and oriented differently to the stellar population and this also supports such an origin.

4. Future Directions

Clearly the nature and very existence of MQGs at $z > 3$, and up to $z = 5$, continues to challenge core ideas in galaxy formation models. It remains a hot topic. My 'to do' list for the community is extensive and challenging. Key priorities include:

(1) We need to firmly establish the number density from $z = 3$ to the maximum possible redshift, this requires wider area JWST surveys in multiple filters (to ensure good photometric identification) with full spectroscopic confirmation. Surveys should aim to reach a square degree on the sky, as a bonus this will allow definitive clustering measurements.
(2) We need to establish the stellar mass to halo mass relation ideally via multiple complementary techniques.
(3) Further high resolution spectroscopy from JWST can look at elemental abundance patterns; chemical evolution models can then further constrain formation histories. How exotic are the stellar populations compared to the modern universe? One can think of $z > 3$ MQGs as providing a fossil record as early star formation where nature has helpfully gathered a lot of stars in to a bright object permitting high signal:noise spectroscopy.
(4) Further 2D Integral Field observations from JWST could allow the IMF and mass structure of MQGs to be investigated and related to their stellar populations.
(5) A considerable amount of ALMA time should be invested in studying the molecular gas content, comparing with neutral gas and using this to constrain quenching mechanisms. Ideally molecular gas will be *detected.*
(6) Wider surveys can also find more of the rarer 'early formers' so they can be placed in context and properly compared with simulations. What is the true abundance of MQGs forming at $z > 7$?
(7) It would be interesting to see if the most extreme MQGs (such as the early formers) could be connected to more exotic cosmological or dark matter models (Glazebrook et al. 2024) and perhaps to other 'too much too soon' populations revealed by JWST such as the overabundance of $z > 10$ luminous star forming galaxies and the overabundance of $z > 7$ AGN. (See the short review of early

JWST results from Adamo et al. (2024) for discussion and more references of these). Such exotic models would need to be tested against observations in true hydrodynamical simulations to properly track galaxy assembly.

Finally I would like to end with a challenge to the community. New deep infrared sky surveys (e.g. from Euclid or Roman missions) have the potential to find the first examples of *strongly lensed* MQGs at $z > 3$. None are currently known beyond $z = 2$. If these could be found the lensing amplification in both brightness and spatial resolution would open up precision studies with JWST which could really elucidate the formation of this mysterious population.

Acknowledgements

K. G. acknowledges useful discussions with the majority of the the authors referenced here over the years and would esp. like to thank T. Nanayakkara at Swinburne for his strong collaboration on the JWST science. K. G. also acknowledges support from Australian Research Council Laureate Fellowship FL180100060 for his JWST research and gives his utmost thanks to the NASA and ESA engineers for building such a remarkable telescope.

References

Adamo, A., Atek, H., Bagley, M. B., Baados, E., Barrow, K. S. S., Berg, D. A., Bezanson, R., Brada, M., Brammer, G., Carnall, A. C., Chisholm, J., Coe, D., Dayal, P., Eisenstein, D. J., Eldridge, J. J., Ferrara, A., Fujimoto, S., Graaff, A. d., Habouzit, M., Hutchison, T. A., Kartaltepe, J. S., Kassin, S. A., Kriek, M., Labb, I., Maiolino, R., Marques-Chaves, R., Maseda, M. V., Mason, C., Matthee, J., McQuinn, K. B. W., Meynet, G., Naidu, R. P., Oesch, P. A., Pentericci, L., Prez-Gonzlez, P. G., Rigby, J. R., Roberts-Borsani, G., Schaerer, D., Shapley, A. E., Stark, D. P., Stiavelli, M., Strom, A. L., Vanzella, E., Wang, F., Wilkins, S. M., Williams, C. C., Willott, C. J., Wylezalek, D., & Nota, A. 2024, The First Billion Years, According to JWST. *arXiv*, 2405.21054.

Alberts, S., Williams, C. C., Helton, J. M., Suess, K. A., Ji, Z., Shivaei, I., Lyu, J., Rieke, G., Baker, W. M., Bonaventura, N., Bunker, A. J., Carniani, S., Charlot, S., Curtis-Lake, E., D'Eugenio, F., Eisenstein, D. J., de Graaff, A., Hainline, K. N., Hausen, R., Johnson, B. D., Maiolino, R., Parlanti, E., Rieke, M. J., Robertson, B. E., Sun, Y., Tacchella, S., Willmer, C. N. A., & Willott, C. J. 2024, To High Redshift and Low Mass: Exploring the Emergence of Quenched Galaxies and Their Environments at $3 < z < 6$ in the Ultra-deep JADES MIRI F770W Parallel. *ApJ*, 975(1), 85.

Baldry, I. K., Glazebrook, K., & Driver, S. P. 2008, On the galaxy stellar mass function, the mass-metallicity relation and the implied baryonic mass function. *Monthly Notices of the Royal Astronomical Society*, 388(3), 945–959.

Belli, S., Park, M., Davies, R. L., Mendel, J. T., Johnson, B. D., Conroy, C., Benton, C., Bugiani, L., Emami, R., Leja, J., Li, Y., Maheson, G., Mathews, E. P., Naidu, R. P., Nelson, E. J., Tacchella, S., Terrazas, B. A., & Weinberger, R. 2024, Star formation shut down by multiphase gas outflow in a galaxy at a redshift of 2.45. *Nature*, 630(8015), 54–58.

Böker, T., Beck, T. L., Birkmann, S. M., Giardino, G., Keyes, C., Kumari, N., Muzerolle, J., Rawle, T., Zeidler, P., Abul-Huda, Y., Alves de Oliveira, C., Arribas, S., Bechtold, K., Bhatawdekar, R., Bonaventura, N., Bunker, A. J., Cameron, A. J., Carniani, S., Charlot, S., Curti, M., Espinoza, N., Ferruit, P., Franx, M., Jakobsen, P., Karakla, D., López-Caniego, M., Lützgendorf, N., Maiolino, R., Manjavacas, E., Marston, A. P., Moseley, S. H., Ogle, P., Perna, M., Peña-Guerrero, M., Pirzkal, N., Plesha, R., Proffitt, C. R., Rauscher, B. J., Rix, H. W., Rodríguez del Pino, B., Rustamkulov, Z., Sabbi, E., Sing, D. K., Sirianni, M., te Plate, M., Úbeda, L., Wahlgren, G. M., Wislowski, E., Wu, R., & Willott, C. J. 2023, In-orbit Performance of the Near-infrared Spectrograph NIRSpec on the James Webb Space Telescope. *PASP*, 135(1045), 038001.

Brammer, G. B., Whitaker, K. E., van Dokkum, P. G., Marchesini, D., Labbé, I., Franx, M., Kriek, M., Quadri, R. F., Illingworth, G., Lee, K. S., Muzzin, A., & Rudnick, G. 2009, The Dead Sequence: A Clear Bimodality in Galaxy Colors from z = 0 to z = 2.5. *ApJ*, 706(1), L173–L177.

Carnall, A. C., Cullen, F., McLure, R. J., McLeod, D. J., Begley, R., Donnan, C. T., Dunlop, J. S., Shapley, A. E., Rowlands, K., Almaini, O., Arellano-Crdova, K. Z., Barrufet, L., Cimatti, A., Ellis, R. S., Grogin, N. A., Hamadouche, M. L., Illingworth, G. D., Koekemoer, A. M., Leung, H. H., Lovell, C. C., Prez-Gonzlez, P. G., Santini, P., Stanton, T. M., & Wild, V. 2024, The JWST EXCELS survey: too much, too young, too fast? Ultra-massive quiescent galaxies at $3 < z < 5$. *Monthly Notices of the Royal Astronomical Society*, 534(1), 325–348.

Carnall, A. C., McLeod, D. J., McLure, R. J., Dunlop, J. S., Begley, R., Cullen, F., Donnan, C. T., Hamadouche, M. L., Jewell, S. M., Jones, E. W., Pollock, C. L., & Wild, V. 2023,a A surprising abundance of massive quiescent galaxies at $3 < z < 5$ in the first data from JWST CEERS. *Monthly Notices of the Royal Astronomical Society*, 520a(3), 3974–3985.

Carnall, A. C., McLure, R. J., Dunlop, J. S., McLeod, D. J., Wild, V., Cullen, F., Magee, D., Begley, R., Cimatti, A., Donnan, C. T., Hamadouche, M. L., Jewell, S. M., & Walker, S. 2023,b A massive quiescent galaxy at redshift 4.658. *Nature*, 619b(7971), 716–719.

Chabrier, G. 2003, Galactic Stellar and Substellar Initial Mass Function. *PASP*, 115(809), 763–795.

Cimatti, A., Cassata, P., Pozzetti, L., Kurk, J., Mignoli, M., Renzini, A., Daddi, E., Bolzonella, M., Brusa, M., Rodighiero, G., Dickinson, M., Franceschini, A., Zamorani, G., Berta, S., Rosati, P., & Halliday, C. 2008, GMASS ultradeep spectroscopy of galaxies at z ~2. II. Superdense passive galaxies: how did they form and evolve? *Astronomy and Astrophysics*, 482(1), 21–42.

Cimatti, A., Daddi, E., Renzini, A., Cassata, P., Vanzella, E., Pozzetti, L., Cristiani, S., Fontana, A., Rodighiero, G., Mignoli, M., & Zamorani, G. 2004, Old galaxies in the young Universe. *Nature*, 430(6), 184–187.

Daddi, E., Renzini, A., Pirzkal, N., Cimatti, A., Malhotra, S., Stiavelli, M., Xu, C., Pasquali, A., Rhoads, J. E., Brusa, M., Alighieri, S. d. S., Ferguson, H. C., Moustakas, L. A., Panagia, N., & Windhorst, R. A. 2005, Passively Evolving Early-Type Galaxies at $1.4 \lesssim z \lesssim 2.5$ in the Hubble Ultra Deep Field. *The Astrophysical Journal*, 626(2), 680–697.

Damjanov, I., McCarthy, P. J., Abraham, R. G., Glazebrook, K., Yan, H., Mentuch, E., Le Borgne, D., Savaglio, S., Crampton, D., Murowinski, R., Juneau, S., Carlberg, R. G., Jørgensen, I., Roth, K., Chen, H.-W., & Marzke, R. O. 2009, Red Nuggets at $z \sim 1.5$: Compact Passive Galaxies and the Formation of the Kormendy Relation. *ApJ*, 695(1), 101–115.

Dekel, A., Birnboim, Y., Engel, G., Freundlich, J., Goerdt, T., Mumcuoglu, M., Neistein, E., Pichon, C., Teyssier, R., & Zinger, E. 2009, Cold streams in early massive hot haloes as the main mode of galaxy formation. *Nature*, 457(7228), 451–454.

d'Eugenio, F., Prez-Gonzlez, P. G., Maiolino, R., Scholtz, J., Perna, M., Circosta, C., bler, H., Arribas, S., Bker, T., Bunker, A. J., Carniani, S., Charlot, S., Chevallard, J., Cresci, G., Curtis-Lake, E., Jones, G. C., Kumari, N., Lamperti, I., Looser, T. J., Parlanti, E., Rix, H.-W., Robertson, B., Pino, B. R. D., Tacchella, S., Venturi, G., & Willott, C. J. 2024, A fast-rotator post-starburst galaxy quenched by supermassive black-hole feedback at z = 3. *Nature Astronomy*, 8(11), 1443–1456.

Esdaile, J., Glazebrook, K., Labbé, I., Taylor, E., Schreiber, C., Nanayakkara, T., Kacprzak, G. G., Oesch, P. A., Tran, K.-V. H., Papovich, C., Spitler, L., & Straatman, C. M. S. 2021, Consistent Dynamical and Stellar Masses with Potential Light IMF in Massive Quiescent Galaxies at $3 < z < 4$ Using Velocity Dispersions Measurements with MOSFIRE. *ApJ*, 908(2), L35.

Forrest, B., Annunziatella, M., Wilson, G., Marchesini, D., Muzzin, A., Cooper, M. C., Marsan, Z. C., McConachie, I., Chan, J. C. C., Gomez, P., Kado-Fong, E., Barbera, F. L., Labb, I., Lange-Vagle, D., Nantais, J., Nonino, M., Pea, T., Saracco, P., Stefanon, M., & Burg,

R. F. J. v. d. 2020, An Extremely Massive Quiescent Galaxy at z = 3.493: Evidence of Insufficiently Rapid Quenching Mechanisms in Theoretical Models. *The Astrophysical Journal Letters*, 890(1), L1.

Forrest, B., Marsan, Z. C., Annunziatella, M., Wilson, G., Muzzin, A., Marchesini, D., Cooper, M. C., Chan, J. C. C., McConachie, I., Gomez, P., Kado-Fong, E., La Barbera, F., Lange-Vagle, D., Nantais, J., Nonino, M., Saracco, P., Stefanon, M., & van der Burg, R. F. J. 2020, The Massive Ancient Galaxies at $z > 3$ NEar-infrared (MAGAZ3NE) Survey: Confirmation of Extremely Rapid Star Formation and Quenching Timescales for Massive Galaxies in the Early Universe. *ApJ*, 903(1), 47.

Forrest, B., Wilson, G., Muzzin, A., Marchesini, D., Cooper, M. C., Marsan, Z. C., Annunziatella, M., McConachie, I., Zaidi, K., Gomez, P., Stawinski, S. M. U., Chang, W., Lucia, G. d., Barbera, F. L., Lubin, L., Nantais, J., Pea, T., Saracco, P., Surace, J., & Stefanon, M. 2022, MAGAZ3NE: High Stellar Velocity Dispersions for Ultramassive Quiescent Galaxies at $z \gtrsim 3$. *The Astrophysical Journal*, 938(2), 109.

Genel, S., Vogelsberger, M., Springel, V., Sijacki, D., Nelson, D., Snyder, G., Rodriguez-Gomez, V., Torrey, P., & Hernquist, L. 2014, Introducing the Illustris project: the evolution of galaxy populations across cosmic time. *MNRAS*, 445(1), 175–200.

Glazebrook, K., Abraham, R. G., McCarthy, P. J., Savaglio, S., Chen, H.-W., Crampton, D., Murowinski, R., Jorgensen, I., Roth, K., Hook, I., Marzke, R. O., & Carlberg, R. G. 2004, A high abundance of massive galaxies 3-6 billion years after the Big Bang. *Nature*, 430(6), 181–184.

Glazebrook, K., Nanayakkara, T., Schreiber, C., Lagos, C., Kawinwanichakij, L., Jacobs, C., Chittenden, H., Brammer, G., Kacprzak, G. G., Labbe, I., Marchesini, D., Marsan, Z. C., Oesch, P. A., Papovich, C., Remus, R.-S., Tran, K.-V. H., Esdaile, J., & Chandro-Gomez, A. 2024, A massive galaxy that formed its stars at z $\approx$ 11. *Nature*, 628(8007), 277–281.

Glazebrook, K., Schreiber, C., Labbé, I., Nanayakkara, T., Kacprzak, G. G., Oesch, P. A., Papovich, C., Spitler, L. R., Straatman, C. M. S., Tran, K.-V. H., & Yuan, T. 2017, A massive, quiescent galaxy at a redshift of 3.717. *Nature*, 544(7648), 71–74.

Graaff, A. d., Setton, D. J., Brammer, G., Cutler, S., Suess, K. A., Labbe, I., Leja, J., Weibel, A., Maseda, M. V., Whitaker, K. E., Bezanson, R., Boogaard, L. A., Cleri, N. J., Lucia, G. D., Franx, M., Greene, J. E., Hirschmann, M., Matthee, J., McConachie, I., Naidu, R. P., Oesch, P. A., Price, S. H., Rix, H.-W., Valentino, F., Wang, B., & Williams, C. C. 2024, Efficient formation of a massive quiescent galaxy at redshift 4.9. *arXiv*, 2404.05683.

Hon, D. S.-H., Graham, A. W., Davis, B. L., & Marconi, A. 2022, Disc cloaking: Establishing a lower limit to the number density of local compact massive spheroids/bulges and the potential fate of some high-z red nuggets. *Monthly Notices of the Royal Astronomical Society*, 514(3), 3410–3451.

Ito, K., Valentino, F., Brammer, G., Faisst, A. L., Gillman, S., Gómez-Guijarro, C., Gould, K. M. L., Heintz, K. E., Ilbert, O., Jespersen, C. K., Kokorev, V., Kubo, M., Magdis, G. E., McPartland, C. J. R., Onodera, M., Rizzo, F., Tanaka, M., Toft, S., Vijayan, A. P., Weaver, J. R., Whitaker, K. E., & Wright, L. 2024, Size–Stellar Mass Relation and Morphology of Quiescent Galaxies at z $\geq$ 3 in Public JWST Fields. *ApJ*, 964(2), 192.

Kriek, M., van Dokkum, P. G., Franx, M., Illingworth, G. D., Marchesini, D., Quadri, R., Rudnick, G., Taylor, E. N., Förster Schreiber, N. M., Gawiser, E., Labbé, I., Lira, P., & Wuyts, S. 2008, A Near-Infrared Spectroscopic Survey of K-Selected Galaxies at $z \sim 2.3$: Redshifts and Implications for Broadband Photometric Studies. *ApJ*, 677(1), 219–237.

Kurinchi-Vendhan, S., Farcy, M., Hirschmann, M., & Valentino, F. 2024, On the origin of star formation quenching in massive galaxies at z $\gtrsim$ 3 in the cosmological simulations IllustrisTNG. *MNRAS*, 534(4), 3974–3988.

Le Borgne, D., Abraham, R., Daniel, K., McCarthy, P. J., Glazebrook, K., Savaglio, S., Crampton, D., Juneau, S., Carlberg, R. G., Chen, H.-W., Marzke, R. O., Roth, K., Jørgensen, I., & Murowinski, R. 2006, Gemini Deep Deep Survey. VI. Massive Hδ-strong Galaxies at z ~= 1. *ApJ*, 642(1), 48–62.

Magliocchetti, M., Santini, P., Merlin, E., & Pentericci, L. 2023, The clustering properties of high-redshift passive galaxies. *Monthly Notices of the Royal Astronomical Society: Letters*, 526(1), L8–L12.

Martínez-Marín, M., Glazebrook, K., Nanayakkara, T., Jacobs, C., Labbé, I., Kacprzak, G. G., Papovich, C., & Schreiber, C. 2024, The origin of large emission line widths in massive galaxies at redshifts $z \sim$3–4. *MNRAS*, 531(3), 3187–3202.

McCarthy, P. J., Borgne, D. L., Crampton, D., Chen, H.-W., Abraham, R. G., Glazebrook, K., Savaglio, S., Carlberg, R. G., Marzke, R. O., Roth, K., Jorgensen, I., Hook, I., Murowinski, R., & Juneau, S. 2004, Evolved Galaxies at $z > 1.5$ from the Gemini Deep Deep Survey: The Formation Epoch of Massive Stellar Systems. *The Astrophysical Journal*, 614(1), L9–L12.

Mendel, J. T., Beifiori, A., Saglia, R. P., Bender, R., Brammer, G. B., Chan, J., Schreiber, N. M. F., Fossati, M., Galametz, A., Momcheva, I. G., Nelson, E. J., Wilman, D. J., & Wuyts, S. 2020, The Kinematics of Massive Quiescent Galaxies at $1.4 < z < 2.1$: Dark Matter Fractions, IMF Variation, and the Relation to Local Early-type Galaxies. *The Astrophysical Journal*, 899(1), 87.

Merlin, E., Fortuni, F., Torelli, M., Santini, P., Castellano, M., Fontana, A., Grazian, A., Pentericci, L., Pilo, S., & Schmidt, K. B. 2019, Redand dead CANDELS: massive passive galaxies at the dawn of the Universe. *Monthly Notices of the Royal Astronomical Society*, 490(3), 3309–3328.

Nanayakkara, T., Glazebrook, K., Schreiber, C., Chittenden, H., Brammer, G., Esdaile, J., Jacobs, C., Kacprzak, G. G., Kawinwanichakij, L., Kimmig, L. C., Labbe, I., Lagos, C., Marchesini, D., Martìnez-Marìn, M., Marsan, Z. C., Oesch, P. A., Papovich, C., Remus, R.-S., & Tran, K.-V. H. 2024, The formation histories of massive and quiescent galaxies in the $3 < z < 4.5$ Universe. *arXiv*, arXiv:2410.02076.

Pérez-González, P. G., D'Eugenio, F., Rodríguez del Pino, B., Übler, H., Maiolino, R., Arribas, S., Cresci, G., Lamperti, I., Bunker, A. J., Carniani, S., Charlot, S., Willott, C. J., Böker, T., Parlanti, E., Scholtz, J., Venturi, G., Barro, G., Costantin, L., Martín-Navarro, I., Dunlop, J. S., & Magee, D. 2024, Feedback mechanisms stopping the star formation in a pair of massive galaxies in the early Universe. *arXiv*, arXiv:2405.03744.

Riechers, D. A., Bradford, C. M., Clements, D. L., Dowell, C. D., Prez-Fournon, I., Ivison, R. J., Bridge, C., Conley, A., Fu, H., Vieira, J. D., Wardlow, J., Calanog, J., Cooray, A., Hurley, P., Neri, R., Kamenetzky, J., Aguirre, J. E., Altieri, B., Arumugam, V., Benford, D. J., Bethermin, M., Bock, J., Burgarella, D., Cabrera-Lavers, A., Chapman, S. C., Cox, P., Dunlop, J. S., Earle, L., Farrah, D., Ferrero, P., Franceschini, A., Gavazzi, R., Glenn, J., Solares, E. A. G., Gurwell, M. A., Halpern, M., Hatziminaoglou, E., Hyde, A., Ibar, E., Kovcs, A., Krips, M., Lupu, R. E., Maloney, P. R., Martinez-Navajas, P., Matsuhara, H., Murphy, E. J., Naylor, B. J., Nguyen, H. T., Oliver, S. J., Omont, A., Page, M. J., Petitpas, G., Rangwala, N., Roseboom, I. G., Scott, D., Smith, A. J., Staguhn, J. G., Streblyanska, A., Thomson, A. P., Valtchanov, I., Viero, M., Wang, L., Zemcov, M., & Zmuidzinas, J. 2013, A dust-obscured massive maximum-starburst galaxy at a redshift of 6.34. *Nature*, 496(7), 329–333.

Rieke, M. J., Kelly, D. M., Misselt, K., Stansberry, J., Boyer, M., Beatty, T., Egami, E., Florian, M., Greene, T. P., Hainline, K., Leisenring, J., Roellig, T., Schlawin, E., Sun, F., Tinnin, L., Williams, C. C., Willmer, C. N. A., Wilson, D., Clark, C. R., Rohrbach, S., Brooks, B., Canipe, A., Correnti, M., DiFelice, A., Gennaro, M., Girard, J. H., Hartig, G., Hilbert, B., Koekemoer, A. M., Nikolov, N. K., Pirzkal, N., Rest, A., Robberto, M., Sunnquist, B., Telfer, R., Wu, C. R., Ferry, M., Lewis, D., Baum, S., Beichman, C., Doyon, R., Dressler, A., Eisenstein, D. J., Ferrarese, L., Hodapp, K., Horner, S., Jaffe, D. T., Johnstone, D., Krist, J., Martin, P., McCarthy, D. W., Meyer, M., Rieke, G. H., Trauger, J., & Young, E. T. 2023, Performance of NIRCam on JWST in Flight. *PASP*, 135(1044), 028001.

Rigby, J. & et al. 2023, The Science Performance of JWST as Characterized in Commissioning. *PASP*, 135(1046), 048001.

Schreiber, C., Glazebrook, K., Nanayakkara, T., Kacprzak, G. G., Labbé, I., Oesch, P., Yuan, T., Tran, K. V., Papovich, C., Spitler, L., & Straatman, C. 2018,a Near infrared spectroscopy and star-formation histories of $3 \leq z \leq 4$ quiescent galaxies. *A&A*, 618a, A85.

Schreiber, C., Labbé, I., Glazebrook, K., Bekiaris, G., Papovich, C., Costa, T., Elbaz, D., Kacprzak, G. G., Nanayakkara, T., Oesch, P., Pannella, M., Spitler, L., Straatman, C., Tran, K. V., & Wang, T. 2018,b Jekyll & Hyde: quiescence and extreme obscuration in a pair of massive galaxies 1.5 Gyr after the Big Bang. *A&A*, 611b, A22.

Setton, D. J., Khullar, G., Miller, T. B., Bezanson, R., Greene, J. E., Suess, K. A., Whitaker, K. E., Antwi-Danso, J., Atek, H., Brammer, G., Cutler, S. E., Dayal, P., Feldmann, R., Fujimoto, S., Furtak, L. J., Glazebrook, K., Goulding, A. D., Kokorev, V., Labbe, I., Leja, J., Ma, Y., Marchesini, D., Nanayakkara, T., Pan, R., Price, S. H., Siegel, J. C., Shipley, H., Weaver, J. R., Dokkum, P. v., Wang, B., & Williams, C. C. 2024, UNCOVER NIRSpec/PRISM Spectroscopy Unveils Evidence of Early Core Formation in a Massive, Centrally Dusty Quiescent Galaxy at $z_{spec} = 3.97$. *The Astrophysical Journal*, 974(1), 145.

Straatman, C. M. S., Labbe, I., Spitler, L. R., Allen, R., Altieri, B., Brammer, G. B., Dickinson, M., Dokkum, P. v., Inami, H., Glazebrook, K., Kacprzak, G. G., Kawinwanichakij, L., Kelson, D. D., McCarthy, P. J., Mehrtens, N., Monson, A., Murphy, D., Papovich, C., Persson, S. E., Quadri, R., Rees, G., Tomczak, A., Tran, K.-V. H., & Tilvi, V. 2014, A Substantial Population of Massive Quiescent Galaxies at z ~4 from ZFOURGE. *The Astrophysical Journal Letters*, 783(1), L14.

Straatman, C. M. S., Labbe, I., Spitler, L. R., Glazebrook, K., Tomczak, A., Allen, R., Brammer, G. B., Cowley, M., Dokkum, P. v., Kacprzak, G. G., Kawinwanichakij, L., Mehrtens, N., Nanayakkara, T., Papovich, C., Persson, S. E., Quadri, R. F., Rees, G., Tilvi, V., Tran, K.-V. H., & Whitaker, K. E. 2015, The Sizes of Massive Quiescent and Star-forming Galaxies at $z \sim 4$ with ZFOURGE and CANDELS. *The Astrophysical Journal Letters*, 808(1), L29.

Suzuki, T. L., Glazebrook, K., Schreiber, C., Kodama, T., Kacprzak, G. G., Leiton, R., Nanayakkara, T., Oesch, P. A., Papovich, C., Spitler, L., Straatman, C. M. S., Tran, K.-V., & Wang, T. 2022, Low Star Formation Activity and Low Gas Content of Quiescent Galaxies at $z = 3.5$–4.0 Constrained with ALMA. *The Astrophysical Journal*, 936(1), 61.

Tanaka, M., Onodera, M., Shimakawa, R., Ito, K., Kakimoto, T., Kubo, M., Morishita, T., Toft, S., Valentino, F., & Wu, P.-F. 2024, A Protocluster of Massive Quiescent Galaxies at z = 4. *The Astrophysical Journal*, 970(1), 59.

Trujillo, I., Ferré-Mateu, A., Balcells, M., Vazdekis, A., & Sánchez-Blázquez, P. 2014, NGC 1277: A Massive Compact Relic Galaxy in the Nearby Universe. *ApJ*, 780(2), L20.

Trujillo, I., Förster Schreiber, N. M., Rudnick, G., Barden, M., Franx, M., Rix, H.-W., Caldwell, J. A. R., McIntosh, D. H., Toft, S., Häussler, B., Zirm, A., van Dokkum, P. G., Labbé, I., Moorwood, A., Röttgering, H., van der Wel, A., van der Werf, P., & van Starkenburg, L. 2006, The Size Evolution of Galaxies since $z \sim 3$: Combining SDSS, GEMS, and FIRES. *ApJ*, 650(1), 18–41.

Turner, C., Tacchella, S., D'Eugenio, F., Carniani, S., Curti, M., Glazebrook, K., Johnson, B. D., Lim, S., Looser, T., Maiolino, R., Nanayakkara, T., & Wan, J. T. 2024, Age-dating early quiescent galaxies: high star-formation efficiency, but consistent with direct, higher-redshift observations. *arXiv*, arXiv:2410.05377.

Valentino, F., Tanaka, M., Davidzon, I., Toft, S., Gmez-Guijarro, C., Stockmann, M., Onodera, M., Brammer, G., Ceverino, D., Faisst, A. L., Gallazzi, A., Hayward, C. C., Ilbert, O., Kubo, M., Magdis, G. E., Selsing, J., Shimakawa, R., Sparre, M., Steinhardt, C., Yabe, K., & Zabl, J. 2020, Quiescent Galaxies 1.5 Billion Years after the Big Bang and Their Progenitors. *The Astrophysical Journal*, 889(2), 93.

van Dokkum, P. G., Franx, M., Kriek, M., Holden, B., Illingworth, G. D., Magee, D., Bouwens, R., Marchesini, D., Quadri, R., Rudnick, G., Taylor, E. N., & Toft, S. 2008, Confirmation of the Remarkable Compactness of Massive Quiescent Galaxies at $z \sim 2.3$: Early-Type Galaxies Did not Form in a Simple Monolithic Collapse. *ApJ*, 677(1), L5.

van Dokkum, P. G., Whitaker, K. E., Brammer, G., Franx, M., Kriek, M., Labbé, I., Marchesini, D., Quadri, R., Bezanson, R., Illingworth, G. D., Muzzin, A., Rudnick, G., Tal, T., & Wake, D. 2010, The Growth of Massive Galaxies Since z = 2. *ApJ*, 709(2), 1018–1041.

Weller, E. J., Pacucci, F., Ni, Y., Hernquist, L., & Park, M. 2024, Discrepancies Between JWST Observations and Simulations of Quenched Massive Galaxies at $z > 3$: A Comparative Study With IllustrisTNG and ASTRID. *arXiv*, 2406.02664.

Wellons, S., Torrey, P., Ma, C.-P., Rodriguez-Gomez, V., Vogelsberger, M., Kriek, M., Dokkum, P. v., Nelson, E., Genel, S., Pillepich, A., Springel, V., Sijacki, D., Snyder, G., Nelson, D., Sales, L., & Hernquist, L. 2015, The formation of massive, compact galaxies at z = 2 in the Illustris simulation. *Monthly Notices of the Royal Astronomical Society*, 449(1), 361–372.

Williams, R. J., Quadri, R. F., Franx, M., van Dokkum, P., Toft, S., Kriek, M., & Labbé, I. 2010, The Evolving Relations Between Size, Mass, Surface Density, and Star Formation in 3×10^4 Galaxies Since z = 2. *ApJ*, 713(2), 738–750.

Wright, L., Whitaker, K. E., Weaver, J. R., Cutler, S. E., Wang, B., Carnall, A., Suess, K. A., Bezanson, R., Nelson, E., Miller, T. B., Ito, K., & Valentino, F. 2024, Remarkably Compact Quiescent Candidates at $3 < z < 5$ in JWST-CEERS. *The Astrophysical Journal Letters*, 964(1), L10.

Xiao, M., Oesch, P. A., Elbaz, D., Bing, L., Nelson, E. J., Weibel, A., Illingworth, G. D., Dokkum, P. v., Naidu, R. P., Daddi, E., Bouwens, R. J., Matthee, J., Wuyts, S., Chisholm, J., Brammer, G., Dickinson, M., Magnelli, B., Leroy, L., Schaerer, D., Herard-Demanche, T., Lim, S., Barrufet, L., Endsley, R., Fudamoto, Y., Gmez-Guijarro, C., Gottumukkala, R., Labb, I., Magee, D., Marchesini, D., Maseda, M., Qin, Y., Reddy, N. A., Shapley, A., Shivaei, I., Shuntov, M., Stefanon, M., Whitaker, K. E., & Wyithe, J. S. B. 2024, Accelerated formation of ultra-massive galaxies in the first billion years. *Nature*, 635(8038), 311–315.

Zolotov, A., Dekel, A., Mandelker, N., Tweed, D., Inoue, S., DeGraf, C., Ceverino, D., Primack, J. R., Barro, G., & Faber, S. M. 2015, Compaction and quenching of high-z galaxies in cosmological simulations: blue and red nuggets. *MNRAS*, 450(3), 2327–2353.

The First Chapters of Our Cosmic History with JWST
Proceedings IAU Symposium No. 391, 2025
T. Nanayakkara & M. Maseda, eds.
doi:10.1017/S1743921324002485

Catching Galaxies Napping : Star Formation Histories of Post-Starburst Galaxies at Redshifts >4

Gourav Khullar

Department of Astronomy, University of Washington, 3910 15th Ave NE, Seattle, WA 98105, USA
email: gkhullar@uw.edu

Abstract. Through the discovery and characterization of three mini-quenched galaxies at $z = 4 - 6$ in the UNCOVER survey (JWST observations of Abell 2744; PIs Bezanson, Labbe), we discuss the diversity of star formation histories (SFHs) and future prospects for the study of "napping" systems. We find these galaxies with their characteristic features – bright rest-frame UV flux, a weak Balmer/4000Å break, and weak or absent emission lines – corresponding to residual star formation and presence of old stellar populations. With the use of flexible non-parametric SFH models via `Prospector`, the specific star formation rates of these mini-quenched galaxies place them below the star formation main sequence. The posterior distributions of the time-averaged SFRs show that these systems are outliers in the SFR_{10}/SFR_{100} parameter space. We demonstrate that this class of galaxies can best be identified and characterized spectroscopically due to the challenge of photometrically capturing the UV flux, weak Balmer/4000Å break, and weak emission lines simultaneously.

Keywords. quiescent galaxies, gravitational lensing, *JWST*, star formation histories, SED fitting

---

1. Introduction

At high redshifts – $z > 4$, when the Universe was $\sim$ 1 Gyr old or younger – galaxy star formation histories (SFHs) are expected to transition from "stochastic" (or bursty) to "secular" ((Dayal et al. 2014; Caplar and Tacchella 2019; Faisst and Morishita 2024; Ciesla et al. 2024; Dome et al. 2024), leading to the establishment of the star forming main sequence (SFMS) at lower redshifts ($z < 3 - 4$; Leja et al. 2020) (Simulations expect this transition to be primarily caused by the deepening of gravitational potentials such that they withstand feedback-based ejection of star-forming gas). Studies like Ciesla et al. (2024) see a transition mass of $\log_{10}(M_*/M_\odot) < 9$ and $z \sim 9$ where SFH moves from stochastic to secular in the early Universe.

Our ability to characterize and identify the bursty (rising or falling) phase is a function of wavelength coverage and spectral resolution of the observations, as well as the modeling techniques that constrain and quantify "burstiness" in a galaxy's SFH e.g., data that simultaneously constrains emission lines that are SF indicators, rest-frame UV flux, Balmer breaks and absorption features probing recent quenching.

JWST has completely opened up the parameter space of robustly observing these stochastic SFHs, and especially "mini-quenched"/napping galaxies – e.g., Looser et al. (2023b), Looser et al. (2023a), Strait et al. (2023), Cole et al. (2023); Trussler et al. (2024) – with multiple spectroscopically-confirmed discoveries in the past year. These nappers – most likely B-type post starburst galaxies (PSBs) – ought to contain residual or low amounts of SF, as well as old stellar populations; in galaxy spectra, these features are characterized by bright rest-frame UV flux, a weak Balmer/4000Å break, and weak or absent emission lines. But, despite the success of these serendipitous discoveries, much still needs to be learned about the details of these systems. Firstly, nappers are observed to be few and far between, in the broader landscape of spectroscopically observed galaxies with bursty SFHs. Secondly, building a sample of such systems photometrically is a challenge – existing color-color diagnostics do not constrain weak Balmer breaks (Antwi-Danso et al. 2022; Long et al. 2024), requiring unique cuts/rest-frame synthetic filters to constrain the combination of spectral features seen in napping galaxies.

A possible set of solutions is given by the wavelength coverage and sensitivity of JWST NIRSpec/PRISM mode, and the depths achieved with NIRCam imaging in concert with boosts from strong gravitational lensing through lensing field surveys. The UNCOVER – Ultradeep NIRSpec and NIRCam ObserVations before the Epoch of Reionization–Treasury Survey (PIs: Bezanson and Labbe; Bezanson et al. (2022)) is designed to extend the search for quiescent galaxies behind the massive galaxy cluster Abell 2744 (A2744) to unprecedented low masses and high redshifts, i.e., $<10^9 M_\odot$ to $z = 9$. In this work, we present the discovery and characterization of a small sample of mini-quenched galaxies at redshifts 4-6, discovered in the spectroscopic survey component of UNCOVER. This sample widens the mass and redshift range where these galaxies are observed, and doubles the number of spectroscopically confirmed "napping" systems close to the epoch of reionization. This is a part of a broader study to identify quenched systems at redshifts greater than 3, and isolate and study napping galaxies, towards constraining the duty cycle of galaxy star formation and quenching.

2. Data and Sample Selection

We use *JWST*/NIRCam (Rieke et al. 2022) UNCOVER Survey (Bezanson et al. 2022) data and ancillary datasets in Abell 2744 from *HST* (Lotz et al. 2016; Steinhardt et al. 2020; Mahler et al. 2018)) and other *JWST* programs – GLASS (Treu et al. 2022), DDT observations (#2756, PI: Chen). All spectroscopic targets were observed with the Micro-Shutter Assembly (MSA) follow-up of the UNCOVER/*JWST* field Abell 2744 (Bezanson et al. 2022); The spectroscopic experimental design and reductions are laid out in detail in Price et al. (2024, in prep). We use the latest version (`v1.1`) of the Furtak et al. (2022) parametric strong lensing model of Abell 2744; we use magnifications as a multiplicative correction factor.

Here, we visually inspect all 680 observed 1D co-added spectra in both F_ν and F_λ space to identify spectral signatures key to mini-quenched galaxies observed in published studies (Looser et al. 2023b,a; Cole et al. 2023; Strait et al. 2023):

(1) strong upturn in rest-frame UV flux.
(2) both a Lyman-break (912Å) and a weak but present Balmer break (3650Å).
(3) weak or absent Hα emission.

Figure 1 shows an example *JWST*/NIRSpec PRISM spectrum of a in this study at ($z = 5.72$). From this sample, we design a quantitative criterion to select/isolate these galaxies from star forming and massive quiescent systems, as well as to robustly test the detectability of napping galaxies in photometric data (these do not isolate themselves in classic UVJ and *ugi* color-color spaces). From each spectrum, we construct a multi-dimensional rest-frame color space that robustly identifies and separates the UNCOVER napper sample at $4 < z < 7$ from other galaxies. We construct the following TopHat-shaped synthetic photometric filters:

(1) **UV1** : 1600–2100 Å ; sampling the far-UV,
(2) **UV2** : 3400–3600 Å ; sampling the near-UV/blue-ward of the Balmer break,

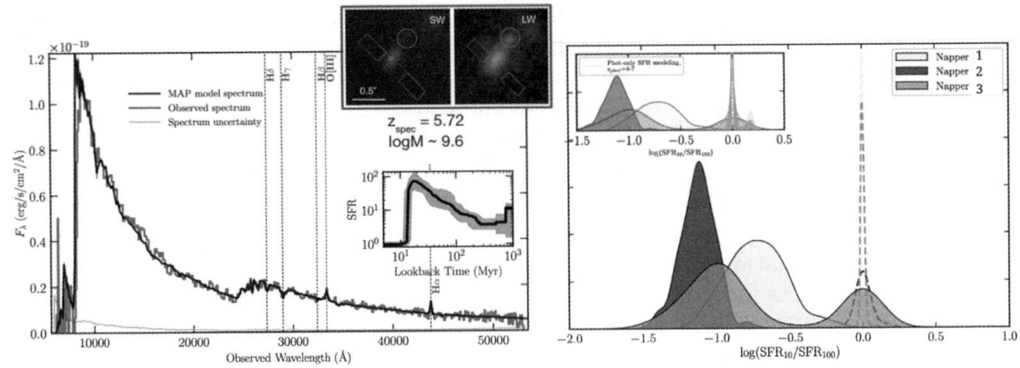

Figure 1. **(Left)** *JWST* NIRSpec PRISM spectrum of the z = 5.72, $\log_{10}(M_*/M_\odot)\sim 9.6$ UNCOVER napper in this work (brown), uncertainties (orange) and the MAP model (black) fit to the spectrophotometric data. The spectrum is characterized by bright rest-frame UV flux, a weak Balmer/4000Å break, and weak or absent emission (as marked by vertical dotted lines). Also depicted is the short- and long- wavelength RGB image of the napper galaxy, as well as the inferred (delensed) posterior distribution of the SFH as a function of lookback time; the median SFH is plotted in bold black. The galaxy has formed the majority of its mass at older times, underwent a quenching episode ~ 10-30 Myr ago, and has sub-solar metallicity. **(Right)** Posterior distributions of the burstiness parameter $\log(\mathrm{SFR}_{10}/\mathrm{SFR}_{100})$ for a random sample of UNCOVER galaxies (photometry-only fits with medium-bands, dotted, and broad-band fits in the inset plot in gold) and mini-quenched/napping galaxies at $4 < z < 6$ (filled regions, spectrophotometric fits) from this study. The mini-quenched galaxies are significant outliers in this parameter space; flexible star formation history modeling constrains the burstiness parameter uncertainties reliably, including bimodal solutions.

(3) **Bal** : 4150–4250 Å ; sampling a non-absorption region of a spectrum redwards of the Balmer break,
(4) **OpRed** : 5300-5500 Å ; sampling redder optical wavelengths, and
(5) **Ha** : 6400-6700 Å ; sampling Hα emission.

We isolate nappers (which are effectively O/B-type PSBs) with the following rest-frame color-cuts:

$$\begin{gathered}(\ \mathbf{UV1}-\mathbf{UV2} < 0.2\ )\ \wedge\ (\ \mathbf{UV2}-\mathbf{Bal} > 0.2\ )\ \wedge \\ (\ \mathbf{Bal}-\mathbf{OpRed} > -0.3\ )\ \wedge\ (\ \mathbf{Opred}-\mathbf{Ha} < 0.6\ )\end{gathered} \tag{1}$$

2.1. *Modeling Star Formation Histories with* `Prospector`

We conduct stellar population synthesis (SPS) modeling via Bayesian SED fitting and generate a star formation history for each system, to conclusively demonstrate the presence/absence of quenching in these galaxies at/near the epoch of observation. For both the spectrophotometric modeling of mini-quenched galaxies in the UNCOVER survey, as well as modeling the $z > 4$ UNCOVER galaxies (for which we only have photometry), we utilize flexible star formation history modeling with the SED fitting framework `prospector` (Johnson and Leja 2017; Leja et al. 2017; Johnson et al. 2021). `Prospector` uses the Flexible Stellar Population Synthesis (FSPS) stellar population synthesis models (Conroy et al. 2009; Conroy and Gunn 2010), the MILES spectral library (Sánchez-Blázquez et al. 2006; Falcón-Barroso et al. 2011), and the MIST isochrones (Choi et al. 2016; Dotter 2016).

We use non-parametric SFHs and implement a flexible age bin model, which utilizes fixed time/age bins at early times, flexible bins that each form the same amount of total stellar mass at intermediate times, and a final age bin with a flexible age boundary (closest to the epoch of observation); this model is optimized to constrain mass assembly in recently quenched galaxies (see Suess et al. 2022b; Setton et al. 2023). This scheme is designed to recover quenching timescales and burst mass fractions with spectro-photometric data (Suess et al. 2022b,a). Here, we describe the model parameter choices pertaining to this work. In our fiducial model, we define the 2 fixed time bins at early times, such that they sample 40% of the age of the Universe ($0.4 \times t_{univ}$), while simultaneously fitting for the redshift, z_{spec}. The youngest age bin is sampled in the analysis in the range $[0, 0.3 \times (t_{univ})]$, as well as the flexible age bins. Moreover, we fit for redshift by sampling the range [0.3, 12].

We assume a Chabrier (2003) initial mass function. We adopt the Kriek and Conroy (2013) dust law with A_v and dust index as free parameters, with doubled attenuation around young ($< 10^7$ yr old) stars following Wild et al. (2020); Suess et al. (2022a); Setton et al. (2023). Finally, we also marginalize over both stellar and gas-phase metallicity (independent of each other). We use the `dynesty` dynamic nested sampling package (Speagle 2020) to sample the posterior distributions. We estimate de-lensed SFRs and stellar masses for the sample using the posterior distributions of the non-parametric star formation history, the total stellar mass formed (corrected for mass loss due to stellar winds, supernovae, etc.) and the multiplicative lensing magnification.

3. Results

The SFH of the example UNCOVER napper is representative of a mini-quenched/napping galaxy, with its constrained star formation history depicting mass build up at $z > 10$, a rapid rise in star formation activity between z$\sim 6-10$, and quenching that began ~20 Myr before the epoch of observation. It is a galaxy that is the most massive quenched system in our sample (logM ($M_\odot$) $= 9.64^{+0.03}_{-0.08}$), with a current star formation rate (SFR) $= 0.65^{+0.13}_{-0.17}$ $M_\odot$ yr^{-1} and specific star formation rate (sSFR) $= -9.81^{+0.06}_{-0.10}$ yr^{-1}, safely placing it in the quenched territory relative to the star forming main sequence (SFMS).

After the serendipitous discovery of GS-z7-01-QU ($z = 7.3$) by Looser et al. (2023b), Looser et al. (2023a) conducted the first spectral search for bursty SFHs and mini-quenching at $z > 4$ in the JADES survey, with `ppxf` (Cappellari 2023) spectral fitting and a library of simple stellar population SED templates. They find significant evidence for bursty SFHs at high redshifts, as well as two galaxies in a "lull" phase – GS-z7-01-QU at $z = 7.3$ ($\log(M_\star/M_\odot) \sim 8.7$), and another at $z = 4.5$ ($\log(M_\star/M_\odot) \sim 7.8$). Strait et al. (2023) showcase the study of another post-starburst system, MACS0417-z5PSB at $z = 5.2$, having a substantially lower stellar mass ($\log(M_\star/M_\odot)$ ~7.6) compared with the characteristic mass at these redshifts (Weaver et al. 2023).

It is worth noting that so far, the post-starburst and quiescent galaxies reported at $z_{spec} > 4$ in *JWST* literature – Strait et al. (2023), Looser et al. (2023b),Looser et al. (2023a)– are low mass objects relative to our sample, either confirming an extension of the mass regime of mini-quenched galaxies in the early Universe, or a unique discovered channel of a quenched phase in stochastic SFH galaxies at $z > 4$. Mini-quenched galaxy candidates have been reported in photometric studies also, albeit without any spectroscopic confirmation currently. Cole et al. (2023) characterize the SFR-M_* relation in CEERS survey galaxies at $4.5 < z < 12$ with the SPS modeling code `BAGPIPES` and find evidence for low-SFR/PSB systems at $\log(M_\star/M_\odot) > 9$. Trussler et al. (2024) demonstrate a photometric search method to find nappers at $< z < 8$, that make up 10-35% of the galaxy population in the JADES survey at $\log(M_\star/M_\odot) \sim$ 8-9 (measured with `BAGPIPES`). Spectroscopic redshifts are yet to be confirmed for these galaxies.

We also estimate the "burstiness" parameter (as seen in literature; see Looser et al. (2023a) for examples) $\log(\mathrm{SFR}_{10}/\mathrm{SFR}_{100})$ for nappers, both via flexible SFH fitting of broad band photometry-only data (dotted kernel density estimates; KDEs) and spectrophotometric data (solid filled KDEs) – see right panel of Figure 1. It is clear that nappers – expected to be outliers in this space – are robustly constrained with joint fits to photometry and spectroscopy, because of the necessary observations of weak Balmer breaks and weak/absent Hα emission. Photometry-only fits to the nappers' observations washes out the above signal, where the probability of retrieving $\log(\mathrm{SFR}_{10}/\mathrm{SFR}_{100})$ < -0.5 is extremely low. It is worth noting that we see bimodal solutions in the burstiness parameter due to the relatively uncertain spectrophotometric fit SFH solution for the galaxy 51660, despite which we confidently place this system as quenched in its current phase.

In the inset of right panel in Figure 1, we compare the nappers' burstiness posteriors (based on joint spectroscopy and photometry fits) to a randomly sampled set of posteriors of photometry-only fits of all UNCOVER galaxies at $4 < z < 7$ (photometry-only fits, gold), demonstrating that broad-band photometry is unable to discover outliers in this parameter space. Note that photometry-only fits are able to isolate a distinct population of systems at $\log(\mathrm{SFR}_{10}/\mathrm{SFR}_{100}) > 0.0$ (peak at ~ 0.2), suggesting that recent starbursts are captured by broad-band SED modeling, while averaging out the signal from galaxies where SFR is dropping rapidly, i.e. $\log(\mathrm{SFR}_{10}/\mathrm{SFR}_{100}) < 0.0$.

Are these galaxies actually "napping", or in the early stages of permanent quiescence? There are several arguments to consider. First, the current set of spectroscopically confirmed nappers from *JWST* are considered low-mass systems ($\log(M_\star/M_\odot)$ $\sim 7-9$), to distinguish them from permanently quenched massive galaxies that are expected to evolve passively (Setton et al. 2024; de Graaff et al. 2024; Marchesini et al. 2022; Nanayakkara et al. 2022). These galaxies – even sometimes labeled as quiescent dwarf galaxies in or near the epoch of reionization (EOR) – are expected to sit in shallow potential wells, where star-forming gas is expelled with ease by merger-driven starbursts, moderate AGN feedback or other galactic winds.

Second, we need to consider the rarity of nappers at $z > 4$. In simulations, the number density of nappers is heavily disputed currently – In IllustrisTNG and VELA, (Dome et al. 2024) find these to be rare ($\sim$1% at z$>$4), while Gelli et al. (2023) find these to be far more common ($>$30% at $z > 6$). Therefore, since we discover three post-starburst galaxies in UNCOVER at $z = 4-6$ with an order of magnitude range of stellar masses in a supposedly transitory period in the Universe after the epoch of reionization, their "napper" candidacy cannot be disputed based on this comparison alone.

Third, observationally, we anticipate that "permanently quenched" systems must likely reside in the most massive dark matter halos at a given redshift Hopkins et al. (2014). At redshifts 4-5, we have spectroscopically confirmed several massive quenched galaxies (Setton et al. 2024; de Graaff et al. 2024; Weibel et al. 2024) with $\log(M_\star/M_\odot)$ > 9. The absence of galaxies that have been quiescent for longer times (and hence have stronger Balmer breaks spectroscopically) at a given redshift and stellar mass threshold likely pushes an observed galaxy from a temporarily quenched phase to potentially a "quenched" system. Conversely, for a galaxy to be mini-quenched (and hence likely to have its SF rejuvenated), it has to have a stellar mass that is at least lower than that of the most massive galaxies that have been spectroscopically characterized at that epoch.

Finally, in theory, the true constraints on stochastic phases of SF would emerge from directly analysing fluctuations in galaxy SFHs. Stochastic modeling of SFHs has recently been introduced as a framework in simulations – e.g., Caplar and Tacchella (2019) – where star formation regulator models of galaxy formation encode small-scale processes related to the life-cycle of giant molecular clouds (GMCs) within galaxies, to processes operating on larger scales, such as the accretion of gas onto galaxies, and major mergers that cause mass assembly and SF activity disruption (Tacchella et al. 2020; Iyer et al.

2024). A promising methodology to observe these fluctuations in simulations is a power spectral density (PSD) analysis, currently widely used in AGN studies. Iyer et al. (2024) use `densebasis`, an SPS Modeling code constructing SFHs via Gaussian Processes (GPs), theoretical SFHs with added stochasticity and toy models of high-redshift galaxy evolution, to demonstrate that PSDs in high-redshift galaxies can be evaluated with some robustness.

4. Discussion and Questions

Adam Carnall: It looked like at least one of your objects is significantly more massive than previously identified "napping" galaxies? How do the masses compare in general and does your finding of relatively massive ones in such a small lensed volume pose questions for current models of how this phase occurs?

G.K.: The logM$\sim$9.5 system is indeed a tricky one, and would technically sit on the boundary of what folks are colloquially labeling a "massive QG" (The other two are logM$\sim$8.5ish, and consistent with scenarios surrounding feedback intermittently disrupting gas flow in weaker potential wells). The most straightforward explanation for the massive system is perhaps a one-time merger driven quenching episode, as opposed to it being in a bursty "lull" phase —but it's only one massive PSB in a small volume so we are cautious here. It is not surprising that we are only seeing the most massive unique ones - lack of emission lines means the more massive PSBs make the observability cut.

Adam Carnall: That sounds very plausible to me, I did wonder if the stochastic star formation stuff that people usually invoke would struggle to explain something at logM $\sim$ 9.5 - I get the impression this is right on the edge of the reasonable range.

Chris Hayward: In FIRE, the mass at which galaxies transition from bursty to time-steady depends on redshift. At $z = 0$, it's true that galaxies of mass logM $\sim$ 9.5 are not very bursty, but at e.g. $z = 2$, they are. We discuss this in detail in Sparre et al. (2017). Unfortunately we only looked at $z = 0$ and 2 there, and in recent work on this topic, we don't have any plots that show this as cleanly.

Joel Leja: The best argument for "napping/mini-quenching" in my opinion is the absence of galaxies with stronger Balmer breaks (i.e. quiescent for longer times) at that mass - to my knowledge that still holds true at logM $\sim$ 9.5?

G.K.: Joel's argument is a solid one —at logM$\sim$9.5 and $z > 5.5$, we don't really have any other spectroscopically confirmed galaxy with "older" stellar populations. There may be hints of such systems in some phot-only modeling papers (but i would again be cautious).

References

Antwi-Danso, J., Papovich, C., Leja, J., Marchesini, D., Marsan, Z. C., Martis, N. S., Labbé, I., Muzzin, A., Glazebrook, K., Straatman, C. M. S., & Tran, K.-V. H. 2022, Beyond UVJ: Color Selection of Galaxies in the JWST Era. *arXiv e-prints*, arXiv:2207.07170.

Bezanson, R., Labbe, I., Whitaker, K. E., Leja, J., Price, S. H., Franx, M., Brammer, G., Marchesini, D., Zitrin, A., Wang, B., Weaver, J. R., Furtak, L. J., Atek, H., Coe, D., Cutler, S. E., Dayal, P., van Dokkum, P., Feldmann, R., Forster Schreiber, N., Fujimoto, S., Geha, M., Glazebrook, K., de Graaff, A., Greene, J. E., Juneau, S., Kassin, S., Kriek, M., Khullar, G., Maseda, M., Mowla, L. A., Muzzin, A., Nanayakkara, T., Nelson, E. J., Oesch, P. A., Pacifici, C., Pan, R., Papovich, C., Setton, D., Shapley, A. E., Smit, R., Stefanon, M., Taylor, E. N., & Williams, C. C. 2022, The JWST UNCOVER Treasury survey: Ultradeep NIRSpec and NIRCam ObserVations before the Epoch of Reionization. *arXiv e-prints*, arXiv:2212.04026.

Caplar, N. & Tacchella, S. 2019, Stochastic modelling of star-formation histories I: the scatter of the star-forming main sequence. *Monthly Notices of the R.A.S.*, 487(3), 3845–3869.

Cappellari, M. 2023, Full spectrum fitting with photometry in PPXF: stellar population versus dynamical masses, non-parametric star formation history and metallicity for 3200 LEGA-C galaxies at redshift $z \approx 0.8$. *Monthly Notices of the R.A.S.*, 526(3), 3273–3300.

Chabrier, G. 2003, Galactic Stellar and Substellar Initial Mass Function. *Publications of the Astronomical Society of Pacific*, 115(809), 763–795.

Choi, J., Dotter, A., Conroy, C., Cantiello, M., Paxton, B., & Johnson, B. D. 2016, Mesa Isochrones and Stellar Tracks (MIST). I. Solar-scaled Models. *Astrophysical Journal*, 823(2), 102.

Ciesla, L., Elbaz, D., Ilbert, O., Buat, V., Magnelli, B., Narayanan, D., Daddi, E., Gómez-Guijarro, C., & Arango-Toro, R. 2024, Identification of a transition from stochastic to secular star formation around z = 9 with JWST. *Astronomy & Astrophysics*, 686, A128.

Cole, J. W., Papovich, C., Finkelstein, S. L., Bagley, M. B., Dickinson, M., Iyer, K. G., Yung, L. Y. A., Ciesla, L., Amorin, R. O., Arrabal Haro, P., Bhatawdekar, R., Calabro, A., Cleri, N. J., de la Vega, A., Dekel, A., Endsley, R., Gawiser, E., Giavalisco, M., Hathi, N. P., Hirschmann, M., Holwerda, B. W., Kartaltepe, J. S., Koekemoer, A. M., Lucas, R. A., Mascia, S., Mobasher, B., Perez-Gonzalez, P. G., Rodighiero, G., Ronayne, K., Tachhella, S., Weiner, B. J., & Wilkins, S. M. 2023, CEERS: Increasing Scatter along the Star-Forming Main Sequence Indicates Early Galaxies Form in Bursts. *arXiv e-prints*, arXiv:2312.10152.

Conroy, C. & Gunn, J. E. 2010, The Propagation of Uncertainties in Stellar Population Synthesis Modeling. III. Model Calibration, Comparison, and Evaluation. *Astrophysical Journal*, 712(2), 833–857.

Conroy, C., Gunn, J. E., & White, M. 2009, The Propagation of Uncertainties in Stellar Population Synthesis Modeling. I. The Relevance of Uncertain Aspects of Stellar Evolution and the Initial Mass Function to the Derived Physical Properties of Galaxies. *Astrophysical Journal*, 699(1), 486–506.

Dayal, P., Ferrara, A., Dunlop, J. S., & Pacucci, F. 2014, Essential physics of early galaxy formation. *Monthly Notices of the R.A.S.*, 445(3), 2545–2557.

de Graaff, A., Setton, D. J., Brammer, G., Cutler, S., Suess, K. A., Labbe, I., Leja, J., Weibel, A., Maseda, M. V., Whitaker, K. E., Bezanson, R., Boogaard, L. A., Cleri, N. J., De Lucia, G., Franx, M., Greene, J. E., Hirschmann, M., Matthee, J., McConachie, I., Naidu, R. P., Oesch, P. A., Price, S. H., Rix, H.-W., Valentino, F., Wang, B., & Williams, C. C. 2024, Efficient formation of a massive quiescent galaxy at redshift 4.9. *arXiv e-prints*, arXiv:2404.05683.

Dome, T., Tacchella, S., Fialkov, A., Ceverino, D., Dekel, A., Ginzburg, O., Lapiner, S., & Looser, T. J. 2024, Mini-quenching of z = 4-8 galaxies by bursty star formation. *Monthly Notices of the R.A.S.*, 527(2), 2139–2151.

Dotter, A. 2016, MESA Isochrones and Stellar Tracks (MIST) 0: Methods for the Construction of Stellar Isochrones. *Astrophysical Journal Supplement*, 222(1), 8.

Faisst, A. L. & Morishita, T. 2024, Dead or Alive? How Bursty Star Formation and Patchy Dust Can Cause Temporary Quiescence in High-redshift Galaxies. *Astrophysical Journal*, 971(1), 47.

Falcón-Barroso, J., Sánchez-Blázquez, P., Vazdekis, A., Ricciardelli, E., Cardiel, N., Cenarro, A. J., Gorgas, J., & Peletier, R. F. 2011, An updated MILES stellar library and stellar population models. *Astronomy & Astrophysics*, 532, A95.

Furtak, L. J., Shuntov, M., Atek, H., Zitrin, A., Richard, J., Lehnert, M. D., & Chevallard, J. 2022, Constraining the physical properties of the first lensed z 9 - 16 galaxy candidates with JWST. *Monthly Notices of the R.A.S.*,.

Gelli, V., Salvadori, S., Ferrara, A., Pallottini, A., & Carniani, S. 2023, Quiescent Low-mass Galaxies Observed by JWST in the Epoch of Reionization. *Astrophysical Journal Letters*, 954(1), L11.

Hopkins, P. F., Kereš, D., Oñorbe, J., Faucher-Giguère, C.-A., Quataert, E., Murray, N., & Bullock, J. S. 2014, Galaxies on FIRE (Feedback In Realistic Environments): stellar feedback explains cosmologically inefficient star formation. *Monthly Notices of the R.A.S.*, 445(1), 581–603.

Iyer, K. G., Speagle, J. S., Caplar, N., Forbes, J. C., Gawiser, E., Leja, J., & Tacchella, S. 2024, Stochastic Modeling of Star Formation Histories. III. Constraints from Physically Motivated Gaussian Processes. *Astrophysical Journal*, 961(1), 53.

Johnson, B., Foreman-Mackey, D., Sick, J., Leja, J., Byler, N., Walmsley, M., Tollerud, E., Leung, H., & Scott, S. 2021,. dfm/python-fsps: python-fsps v0.4.0.

Johnson, B. & Leja, J. 2017,. Bd-J/Prospector: Initial Release.

Kriek, M. & Conroy, C. 2013, The Dust Attenuation Law in Distant Galaxies: Evidence for Variation with Spectral Type. *Astrophysical Journal Letters*, 775(1), L16.

Leja, J., Johnson, B. D., Conroy, C., van Dokkum, P. G., & Byler, N. 2017, Deriving Physical Properties from Broadband Photometry with Prospector: Description of the Model and a Demonstration of its Accuracy Using 129 Galaxies in the Local Universe. *Astrophysical Journal*, 837(2), 170.

Leja, J., Speagle, J. S., Johnson, B. D., Conroy, C., van Dokkum, P., & Franx, M. 2020, A New Census of the 0.2 ¡ z ¡ 3.0 Universe. I. The Stellar Mass Function. *Astrophysical Journal*, 893(2), 111.

Long, A. S., Antwi-Danso, J., Lambrides, E. L., Lovell, C. C., de la Vega, A., Valentino, F., Zavala, J. A., Casey, C. M., Wilkins, S. M., Yung, L. Y. A., Arrabal Haro, P., Bagley, M. B., Bisigello, L., Chworowsky, K., Cooper, M. C., Cooper, O. R., Cooray, A. R., Croton, D., Dickinson, M., Finkelstein, S. L., Franco, M., Gould, K. M. L., Hirschmann, M., Hutchison, T. A., Kartaltepe, J. S., Kocevski, D. D., Koekemoer, A. M., Lucas, R. A., McKinney, J., Nere, R., Papovich, C., Pérez-González, P. G., Pirzkal, N., & Santini, P. 2024, Efficient NIRCam Selection of Quiescent Galaxies at 3 ¡ z ¡ 6 in CEERS. *Astrophysical Journal*, 970(1), 68.

Looser, T. J., D'Eugenio, F., Maiolino, R., Tacchella, S., Curti, M., Arribas, S., Baker, W. M., Baum, S., Bonaventura, N., Boyett, K., Bunker, A. J., Carniani, S., Charlot, S., Chevallard, J., Curtis-Lake, E., Danhaive, A. L., Eisenstein, D. J., de Graaff, A., Hainline, K., Ji, Z., Johnson, B. D., Kumari, N., Nelson, E., Parlanti, E., Rix, H.-W., Robertson, B., Rodríguez Del Pino, B., Sandles, L., Scholtz, J., Smit, R., Stark, D. P., Übler, H., Williams, C. C., Willott, C., & Witstok, J. 2023,a JADES: Differing assembly histories of galaxies – Observational evidence for bursty SFHs and (mini-)quenching in the first billion years of the Universe. *arXiv e-prints*, arXiv:2306.02470.

Looser, T. J., D'Eugenio, F., Maiolino, R., Witstok, J., Sandles, L., Curtis-Lake, E., Chevallard, J., Tacchella, S., Johnson, B. D., Baker, W. M., Suess, K. A., Carniani, S., Ferruit, P., Arribas, S., Bonaventura, N., Bunker, A. J., Cameron, A. J., Charlot, S., Curti, M., de Graaff, A., Maseda, M. V., Rawle, T., Rix, H.-W., Rodriguez Del Pino, B., Smit, R., Übler, H., Willott, C., Alberts, S., Egami, E., Eisenstein, D. J., Endsley, R., Hausen, R., Rieke, M., Robertson, B., Shivaei, I., Williams, C. C., Boyett, K., Chen, Z., Ji, Z., Jones, G. J., Kumari, N., Nelson, E., Perna, M., Saxena, A., & Scholtz, J. 2023,b Discovery of a quiescent galaxy at z = 7.3. *arXiv e-prints*, arXiv:2302.14155.

Lotz, P. J., Dubois-Felsmann, G. P., Lim, K.-T., Johnson, T., Chandrasekharan, S., Mills, D., Daly, P., Schumacher, G., Delgado, F., Pietrowicz, S., Selvy, B., Sebag, J., Marshall, S., Sundararaman, H., Contaxis, C., Bovill, R., & Jenness, T. LSST control software component design. In Chiozzi, G. & Guzman, J. C., editors, *Software and Cyberinfrastructure for Astronomy IV* 2016, volume 9913 of *Society of Photo-Optical Instrumentation Engineers (SPIE) Conference Series*, 991309.

Mahler, G., Richard, J., Clément, B., Lagattuta, D., Schmidt, K., Patrício, V., Soucail, G., Bacon, R., Pello, R., Bouwens, R., Maseda, M., Martinez, J., Carollo, M., Inami, H., Leclercq, F., & Wisotzki, L. 2018, Strong-lensing analysis of A2744 with MUSE and Hubble Frontier Fields images. *Monthly Notices of the R.A.S.*, 473(1), 663–692.

Marchesini, D., Brammer, G., Morishita, T., Bergamini, P., Wang, X., Bradac, M., Roberts-Borsani, G., Strait, V., Treu, T., Fontana, A., Jones, T., Santini, P., Vulcani, B., Acebron, A., Calabrò, A., Castellano, M., Glazebrook, K., Grillo, C., Mercurio, A., Nanayakkara, T., Rosati, P., Tubthong, C., & Vanzella, E. 2022, Early results from GLASS-JWST. IX: First spectroscopic confirmation of low-mass quiescent galaxies at $z > 2$ with NIRISS. *arXiv e-prints*, arXiv:2207.13625.

Nanayakkara, T., Glazebrook, K., Jacobs, C., Schreiber, C., Brammer, G., Esdaile, J., Kacprzak, G. G., Labbe, I., Lagos, C., Marchesini, D., Marsan, Z. C., Nateghi, H., Oesch, P. A.,

Papovich, C., Remus, R.-S., & Tran, K.-V. H. 2022, A population of faint, old, and massive quiescent galaxies at 3 ¡ z ¡ 4 revealed by JWST NIRSpec Spectroscopy. *arXiv e-prints*, arXiv:2212.11638.

Rieke, M. J., Kelly, D. M., Misselt, K., Stansberry, J., Boyer, M., Beatty, T., Egami, E., Florian, M., Greene, T. P., & Hainline, K. 2022, NIRCam Performance on JWST In Flight. *arXiv e-prints*, arXiv:2212.12069.

Sánchez-Blázquez, P., Peletier, R. F., Jiménez-Vicente, J., Cardiel, N., Cenarro, A. J., Falcón-Barroso, J., Gorgas, J., Selam, S., & Vazdekis, A. 2006, Medium-resolution Isaac Newton Telescope library of empirical spectra. *Monthly Notices of the R.A.S.*, 371(2), 703–718.

Setton, D. J., Dey, B., Khullar, G., Bezanson, R., Newman, J. A., Aguilar, J. N., Ahlen, S., Andrews, B. H., Brooks, D., de la Macorra, A., Dey, A., Eftekharzadeh, S., Font-Ribera, A., A Gontcho, S. G., Kremin, A., Juneau, S., Landriau, M., Meisner, A., Miquel, R., Moustakas, J., Pearl, A., Prada, F., Tarlé, G., Siudek, M., Weaver, B. A., Zhou, Z., & Zou, H. 2023, DESI Survey Validation Spectra Reveal an Increasing Fraction of Recently Quenched Galaxies at z 1. *Astrophysical Journal Letters*, 947(2), L31.

Setton, D. J., Khullar, G., Miller, T. B., Bezanson, R., Greene, J. E., Suess, K. A., Whitaker, K. E., Antwi-Danso, J., Atek, H., Brammer, G., Cutler, S. E., Dayal, P., Feldmann, R., Furtak, L. J., Fujimoto, S., Glazebrook, K., Goulding, A. D., Kokorev, V., Labbe, I., Leja, J., Ma, Y., Marchesini, D., Nanayakkara, T., Pan, R., Price, S. H., Siegel, J. C., Shipley, H., Weaver, J. R., van Dokkum, P., Wang, B., & Williams, C. C. 2024, UNCOVER NIRSpec/PRISM Spectroscopy Unveils Evidence of Early Core Formation in a Massive, Centrally Dusty Quiescent Galaxy at $z_{spec} = 3.97$. *arXiv e-prints*, arXiv:2402.05664.

Sparre, M., Hayward, C. C., Feldmann, R., Faucher-Giguère, C.-A., Muratov, A. L., Kereš, D., & Hopkins, P. F. 2017, (Star)bursts of FIRE: observational signatures of bursty star formation in galaxies. *Monthly Notices of the R.A.S.*, 466(1), 88–104.

Speagle, J. S. 2020, DYNESTY: a dynamic nested sampling package for estimating Bayesian posteriors and evidences. *Monthly Notices of the R.A.S.*, 493(3), 3132–3158.

Steinhardt, C. L., Jauzac, M., Acebron, A., Atek, H., Capak, P., Davidzon, I., Eckert, D., Harvey, D., Koekemoer, A. M., Lagos, C. D. P., Mahler, G., Montes, M., Niemiec, A., Nonino, M., Oesch, P. A., Richard, J., Rodney, S. A., Schaller, M., Sharon, K., Strolger, L.-G., Allingham, J., Amara, A., Bahé, Y., Bœhm, C., Bose, S., Bouwens, R. J., Bradley, L. D., Brammer, G., Broadhurst, T., Cañas, R., Cen, R., Clément, B., Clowe, D., Coe, D., Connor, T., Darvish, B., Diego, J. M., Ebeling, H., Edge, A. C., Egami, E., Ettori, S., Faisst, A. L., Frye, B., Furtak, L. J., Gómez-Guijarro, C., Remolina González, J. D., Gonzalez, A., Graur, O., Gruen, D., Harvey, D., Hensley, H., Hovis-Afflerbach, B., Jablonka, P., Jha, S. W., Jullo, E., Kneib, J.-P., Kokorev, V., Lagattuta, D. J., Limousin, M., von der Linden, A., Linzer, N. B., Lopez, A., Magdis, G. E., Massey, R., Masters, D. C., Maturi, M., McCully, C., McGee, S. L., Meneghetti, M., Mobasher, B., Moustakas, L. A., Murphy, E. J., Natarajan, P., Neyrinck, M., O'Connor, K., Oguri, M., Pagul, A., Rhodes, J., Rich, R. M., Robertson, A., Sereno, M., Shan, H., Smith, G. P., Sneppen, A., Squires, G. K., Tam, S.-I., Tchernin, C., Toft, S., Umetsu, K., Weaver, J. R., van Weeren, R. J., Williams, L. L. R., Wilson, T. J., Yan, L., & Zitrin, A. 2020, The BUFFALO HST Survey. *Astrophysical Journal Supplement*, 247(2), 64.

Strait, V., Brammer, G., Muzzin, A., Desprez, G., Asada, Y., Abraham, R., Bradač, M., Iyer, K. G., Martis, N., Mowla, L., Noirot, G., Sarrouh, G. T. E., Sawicki, M., Willott, C., Gould, K., Grindlay, T., Matharu, J., & Rihtaršič, G. 2023, An Extremely Compact, Low-mass Galaxy on its Way to Quiescence at z = 5.2. *Astrophysical Journal Letters*, 949(2), L23.

Suess, K. A., Kriek, M., Bezanson, R., Greene, J. E., Setton, D., Spilker, J. S., Feldmann, R., Goulding, A. D., Johnson, B. D., Leja, J., Narayanan, D., Hall-Hooper, K., Hunt, Q., Lower, S., & Verrico, M. 2022,a {SQuIGG}ěc{L}{E} : Studying Quenching in Intermediate-z Galaxies-Gas, {Angu}ěc{L}{ar} Momentum, and Evolution. *Astrophysical Journal*, 926a(1), 89.

Suess, K. A., Leja, J., Johnson, B. D., Bezanson, R., Greene, J. E., Kriek, M., Lower, S., Narayanan, D., Setton, D. J., & Spilker, J. S. 2022,b Recovering the Star Formation

Histories of Recently Quenched Galaxies: The Impact of Model and Prior Choices. *Astrophysical Journal*, 935b(2), 146.

Tacchella, S., Forbes, J. C., & Caplar, N. 2020, Stochastic modelling of star-formation histories II: star-formation variability from molecular clouds and gas inflow. *Monthly Notices of the R.A.S.*, 497(1), 698–725.

Treu, T., Roberts-Borsani, G., Bradac, M., Brammer, G., Fontana, A., Henry, A., Mason, C., Morishita, T., Pentericci, L., Wang, X., Acebron, A., Bagley, M., Bergamini, P., Belfiori, D., Bonchi, A., Boyett, K., Boutsia, K., Calabró, A., Caminha, G. B., Castellano, M., Dressler, A., Glazebrook, K., Grillo, C., Jacobs, C., Jones, T., Kelly, P. L., Leethochawalit, N., Malkan, M. A., Marchesini, D., Mascia, S., Mercurio, A., Merlin, E., Nanayakkara, T., Nonino, M., Paris, D., Poggianti, B., Rosati, P., Santini, P., Scarlata, C., Shipley, H. V., Strait, V., Trenti, M., Tubthong, C., Vanzella, E., Vulcani, B., & Yang, L. 2022, The GLASS-JWST Early Release Science Program. I. Survey Design and Release Plans. *Astrophysical Journal*, 935(2), 110.

Trussler, J., Conselice, C., Adams, N., Austin, D., Caruana, J., Harvey, T., Li, Q., Lovell, C., Seeyave, L., Vijayan, A., & Wilkins, S. 2024, Like a candle in the wind: The embers of once aflame, now smouldering galaxies at $5 < z < 8$. *arXiv e-prints*, arXiv:2404.07163.

Weaver, J. R., Davidzon, I., Toft, S., Ilbert, O., McCracken, H. J., Gould, K. M. L., Jespersen, C. K., Steinhardt, C., Lagos, C. D. P., Capak, P. L., Casey, C. M., Chartab, N., Faisst, A. L., Hayward, C. C., Kartaltepe, J. S., Kauffmann, O. B., Koekemoer, A. M., Kokorev, V., Laigle, C., Liu, D., Long, A., Magdis, G. E., McPartland, C. J. R., Milvang-Jensen, B., Mobasher, B., Moneti, A., Peng, Y., Sanders, D. B., Shuntov, M., Sneppen, A., Valentino, F., Zalesky, L., & Zamorani, G. 2023, COSMOS2020: The galaxy stellar mass function. The assembly and star formation cessation of galaxies at 0.2¡ z $\leq$ 7.5. *Astronomy & Astrophysics*, 677, A184.

Weibel, A., de Graaff, A., Setton, D. J., Miller, T. B., Oesch, P. A., Brammer, G., Lagos, C. D. P., Whitaker, K. E., Williams, C. C., Baggen, J. F. W., Bezanson, R., Boogaard, L. A., Cleri, N. J., Greene, J. E., Hirschmann, M., Hviding, R. E., Kuruvanthodi, A., Labbé, I., Leja, J., Maseda, M. V., Matthee, J., McConachie, I., Naidu, R. P., Roberts-Borsani, G., Schaerer, D., Suess, K. A., Valentino, F., van Dokkum, P., & Wang, B. 2024, RUBIES Reveals a Massive Quiescent Galaxy at z = 7.3. *arXiv e-prints*, arXiv:2409.03829.

Wild, V., Taj Aldeen, L., Carnall, A., Maltby, D., Almaini, O., Werle, A., Wilkinson, A., Rowlands, K., Bolzonella, M., Castellano, M., Gargiulo, A., McLure, R., Pentericci, L., & Pozzetti, L. 2020, The star formation histories of z $\sim$ 1 post-starburst galaxies. *Monthly Notices of the R.A.S.*, 494(1), 529–548.

The First Chapters of Our Cosmic History with JWST
Proceedings IAU Symposium No. 391, 2025
T. Nanayakkara & M. Maseda, eds.
doi:10.1017/S1743921325000079

The FENIKS Survey: Spectroscopic Confirmation of Massive Quiescent Galaxies at $z > 3$

Jacqueline Antwi-Danso, on behalf of the FENIKS team

David A. Dunlap Department of Astronomy & Astrophysics, University of Toronto,
50 St George Street, Toronto, ON M5S 3H4, Canada
email: j.antwidanso@utoronto.ca

Abstract. We report spectroscopic confirmation of a $z = 3.757 \pm 0.0011$ massive ($\log(\mathrm{M}_*/M_\odot) = 10.86^{+0.09}_{-0.03}$) quiescent galaxy with 11 hours of Keck/MOSFIRE K-band observations. This galaxy was selected from the FENIKS survey, which uses deep Gemini/*Flamingos-2* $K_b K_r$ imaging optimized for increased sensitivity to galaxies with strong Balmer/4000 Å breaks. Its rest-frame colors are consistent with a ~ 1 Gyr old stellar population. This places it as one of the oldest objects at these redshifts, and challenges the notion that quiescent galaxies at $z > 3$ are all young post-starbursts. The star formation history suggests that its progenitor was in place by $z \sim 4 - 7$, reaching $10^{10.6}\mathrm{M}_\odot$ by $z \sim 5$, which we have confirmed using ultra-deep medium resolution spectroscopy from *JWST*. These results are in line with the rapid formation and quenching timescales reported for similar massive quiescent galaxies at $z > 3$, implying extreme baryonic physics within the first billion years of cosmic history.

Keywords. High-redshift galaxies, Galaxy evolution, Post-starburst galaxies, Quenched galaxies

1. Introduction

Explaining the existence of massive quiescent galaxies at $z > 3$ with stellar masses larger than the present-day Milky Way mass ($\mathrm{M}_* > 10^{10.7}\mathrm{M}_\odot$) is one of the unsolved problems of modern astrophysics. There are two long-standing tensions between observations of high redshift massive quiescent galaxies and theory: 1) cosmological simulations underestimate their number densities by up to an order of magnitude (see Valentino et al. 2023 for a compilation of observational constraints); and 2) observations suggest that they form much *faster* and *earlier* than in simulations (Hartley et al. 2023, Kimmig et al. 2023). In particular, the stellar masses of the oldest quiescent galaxies are strongly constraining on the halo mass function predicted by ΛCDM given its rapid evolution at high redshift (Glazebrook et al. 2024, Carnall et al. 2024). Therefore, robust spectroscopic confirmation and detailed characterization of this population is paramount for constraining galaxy formation models at early times.

However, massive quiescent galaxies have received limited spectroscopic coverage, as this was observationally expensive prior to the launch of the James Webb Space Telescope (*JWST*). In fact, there are less than 65 spectroscopically-confirmed quiescent galaxies at $z > 3$ (Antwi-Danso et al. 2023a). To this end, we present spectroscopy of three massive quiescent candidates at $z \sim 3 - 5$ identified using medium-band ground-based imaging from the FLAMINGOS-2 Extragalactic Near-Infrared K-band Split (FENIKS) survey (Esdaile et al. 2021). We obtained ultra-deep medium-resolution spectroscopy from Keck/MOSFIRE and *JWST*/NIRSpec General Observer (GO) Program 4318 (PI: Antwi-Danso). The Keck/MOSFIRE observations are presented in Antwi-Danso et al. (2023b) and the NIRSpec observations in a forthcoming paper (Antwi-Danso et al. in prep).

2. Identifying Quiescent Galaxies at $z > 3$ with $K_b K_r$ Medium Bands

Detecting massive quiescent galaxies in the first 2 Gyr presents significant observational challenges. Being composed of primarily of evolved stars (A-type and later), they lack strong emission lines, hence can only be detected via their faint stellar continuum. Their red spectra peak in the near-infrared ($1-5\mu$m), a regime prone to high thermal backgrounds from the ground. Pre-*JWST*, ground-based medium band surveys were our best attempts at addressing these systematics, as they provide higher photometric sampling of the Balmer/4000Å breaks of quiescent galaxies, resulting in improved photometric redshifts and lower outlier rates (Whitaker et al. 2011, Straatman et al. 2014). Even with these improvements, the discovery space from the ground has been limited to the brightest quiescent galaxies at $z < 4$, which tend to be young (< 300 Myr). This introduces an age bias into population studies.

For this work, we use imaging from the 170-hr FENIKS survey, which employs K_b ($\lambda_c = 2.0\mu m$) and K_r ($\lambda_c = 2.3\mu m$) filters installed on Gemini/*Flamingos-2*. These filters are sensitive to the Balmer/4000 Å breaks of massive galaxies at $z > 4$, which falls between these bands at $z = 4.2-5.2$. The increased photometric sampling of the $K_b K_r$ filters therefore ensures more accurate photometric-redshift uncertainties of <3%, and also more accurate constraints on galaxy stellar masses from fitting their spectral energy distributions (SEDs; Muzzin et al. 2009). We demonstrated these in our pilot survey (Esdaile et al. 2021). The FENIKS Gemini LLP covers three extragalactic fields (COSMOS, CDFS, and UDS) with ancillary data from ground-based NIR surveys and *HST* (CANDELS/3D-HST (Grogin et al. 2011), UltraVISTA (McCracken et al. 2012), UKIDSS (Lawrence et al. 2007). Our spectroscopic targets were selected from one of our 6.2′ diameter pointings in COSMOS ($\alpha : 10^h 01^m 49.1016^s$, $\delta : +0228'12.08''$). We selected 4 candidate quiescent galaxies at $z = 3-4$ with two color selection techniques: the traditional UVJ diagram (e.g., Williams et al. 2009) and the $(ugi)_s$ diagram (Antwi-Danso et al. 2023a).

3. Spectroscopic Followup with Keck/MOSFIRE & *JWST*/NIRSpec

We observed our candidates with 11 hours of spectroscopy on MOSFIRE, which has a spectral resolution of $R \sim 3620$ in the K-grating ($1.9-2.4\mu$m). We modeled their multi-wavelength photometry from FENIKS and UltraVISTA and their NIR spectra using three SED fitting codes: eazy-py (photometry only), Bagpipes (photometry+spectra), and slinefit (spectra only), each yielding consistent redshifts and other stellar population parameters. The spectrum of one out of four quiescent candidates (COS-126891) reveals strong, Hβ+[OIII] emission with $\mathrm{EW}_0 = 428.5\mathring{A}$. Given that this galaxy has a large stellar mass of log $(M/M_\odot) = 10.22^{+0.01}_{-0.12}$ (after correcting for emission lines), this is consistent with a broad-line AGN. This is line with AGN comprising at least 60% of massive galaxy samples at $z > 3$ (Marsan et al. 2017). We obtained a robust redshift estimate for one out of three other quiescent candidates (COS-128636, $z = 3.757 \pm 0.0011$) via the detection of strong absorption features (Hβ and Hγ). The other two candidates were too faint ($\mathrm{K_s} > 24$ AB) for robust analysis, revealing that secure spectroscopic confirmation of quiescent galaxies at $z > 4$ is only feasible with *JWST*.

We followed up these candidates with medium resolution ($R \sim 1000$) spectroscopy on *JWST*/NIRSpec, with 3 hrs in the G235M grating and 5 hours in the G395M grating, which combined cover a wavelength range of $\lambda = 0.6 - 5.3$ μm (GO 4318; PI: Antwi-Danso). With this, we confirmed the redshift of COS-128636 and obtained redshifts for the other two quiescent candidates. One of them (COS-129098) turned out to be at the same redshift as COS-128636 ($z = 3.757 \pm 0.0011$), suggesting that they are tracing the same large-scale structure. These two galaxies are some of the faintest spectroscopically-confirmed galaxies from the ground ($K_s \sim 23 - 24$), in agreement with recent *JWST* results suggesting that the bias towards the youngest post-starbursts at $z > 3$ is a selection effect.

4. Rapid Stellar Mass Assembly in the First Billion Years

The star-formation history of COS-128636 suggests that it formed the bulk of its stellar mass at $z > 5$. The mass-weighted age derived using its Keck/MOSFIRE spectrum suggests that the galaxy is $\sim 30 - 60\%$ the age of universe at $z \sim 4$ ($400 - 900$ Myr). Preliminary results from Prospector modeling of its NIRspec spectrum, where the longer wavelength coverage targets a wider suite of age- and metallicity-sensitive features, further constrain its age to $600 - 800$ Myr. This is consistent with the rapid formation and quenching timescales reported for similar massive quiescent galaxies at $z > 3$ (e.g., Carnall et al. 2024). These results highlight the efficiency of *JWST* for studying massive quiescent galaxies in the first two billion years. Additionally, they demonstrate that with *JWST*, we can now obtain robust spectroscopy of the faintest quiescent galaxies at high redshift in order to tackle the age bias in these samples.

References

Antwi-Danso, J., Papovich, C., Leja, J., et al. 2023a, *ApJ*, 943, 166
Antwi-Danso, J., Papovich, C., Esdaile, J., et al. 2023b, arXiv e-prints, arXiv:2307.09590
Carnall, A. C., Cullen, F., McLure, R. J., et al. 2024, *MNRAS*, 534, 325
Esdaile, J., Labbé, I., Glazebrook, K., et al. 2021, *AJ*, 162, 225
Glazebrook, K., Nanayakkara, T., Schreiber, C., et al. 2024, *Nature*, 628, 277
Grogin, N. A., Kocevski, D. D., Faber, S. M., et al. 2011, *ApJ*, 197, 35
Hartley, A. I., Nelson, E. J., Suess, K. A., et al. 2023, *MNRAS*, 522, 3138
Kimmig, L. C., Remus, R.-S., Seidel, B., et al. 2023, arXiv e-prints, arXiv:2310.16085
Lawrence, A., Warren, S. J., Almaini, O., et al. 2007, *MNRAS*, 379, 1599
Marsan, Z. C., Marchesini, D., Brammer, G. B., et al. 2017, *ApJ*, 842, 21
McCracken, H. J., Milvang-Jensen, B., Dunlop, J., et al. 2012, *A&A*, 544, A156
Muzzin, A., Marchesini, D., van Dokkum, P. G., et al. 2009, *ApJ*, 701, 1839
Straatman, C. M. S., Labbé, I., Spitler, L. R., et al. 2014, *ApJL*, 783, L14
Valentino, F., Brammer, G., Gould, K. M. L., et al. 2023, *ApJ*, 947, 20
Whitaker, K. E., Labbé, I., van Dokkum, P. G., et al. 2011, *ApJ*, 735, 86
Williams, R. J., Quadri, R. F., Franx, M., van Dokkum, P., & Labbé, I. 2009, *ApJ*, 691, 1879

The First Chapters of Our Cosmic History with JWST
Proceedings IAU Symposium No. 391, 2025
T. Nanayakkara & M. Maseda, eds.
doi:10.1017/S1743921324002497

Thin and Thick Disk Formation Across Cosmic History

Takafumi Tsukui[1,2], Emily Wisnioski[1,2], Joss Bland-Hawthorn[2,3] and Ken Freeman[1,2]

[1]Research School of Astronomy and Astrophysics, Australian National University, Cotter Road, Weston Creek, ACT 2611, Australia
E-mail: tsukuitk23@gmail.com (TT)

[2]ARC Centre of Excellence for All Sky Astrophysics in 3 Dimensions (ASTRO 3D), Australia

[3]Sydney Institute for Astronomy, School of Physics, A28, The University of Sydney, NSW 2006, Australia

Abstract. Using the superb capabilities of JWST, we investigate the structure of thin and thick disks in galaxies beyond the local universe for the first time. We found evidence of sequential formation, where most galaxies initially form a thick disk, followed by the formation of a thin disk. Thin disk formation occurred around 8 Gyr ago in high-mass galaxies, earlier than 4 Gyr ago in low-mass galaxies, aligning with the onset of thin disk formation in the Milky Way. We propose that this downsizing thin disk formation—delayed in low-mass galaxies—can be naturally explained by stability-regulated disk formation by evolving gas fractions. As gas fractions decrease over time, turbulence in the gas disk declines, allowing a thin disk to form. High-mass galaxies, which efficiently convert gas into stars, achieve lower gas fractions, supporting earlier thin disk formation.

Keywords. galaxies: high-redshift; galaxies: kinematics and dynamics; galaxies: structure; galaxies: evolution

---

1. Introduction

Present-day disk galaxies typically exhibit a young thin disk and an old thick disk (e.g., Burstein 1979; Yoachim & Dalcanton 2006). In the Milky way, most thick disk stars are estimated to have formed as early as $z = 4 - 5$, 12 Gyr to 9 Gyr ago (Belokurov & Kravtsov 2022; Conroy et al. 2022), while the thin disk start forming 9 Gyr ago (Wu et al. 2023), coinciding with the last massive merger 8–10 Gyr ago (Vincenzo et al. 2019) and a significant influx of fresh gas (Ciucă et al. 2024). Thick disks likely form from early turbulent gas disks, satellite accretion and dynamical heating of thin disk stars. As gas turbulence decreased, galaxies are able to support thin disk formation. These processes act together over cosmic time. To untangle these processes, it is essential to investigate thin and thick disks across cosmic epochs, providing a temporal perspective. However, thin and thick disks have only been identified in local galaxies.

2. JWST Edge-on sample and the need for two disk components

To address this gap, we (Tsukui et al. 2024) construct an edge-on galaxy sample using flagship JWST imaging programs along with 3D-HST galaxy parameters and examine the disk structure in rest-frame near-IR images (rest-frame $\sim K, H$ bands) by forward modelling a 3D density distribution with a radially exponential and vertically sech^2

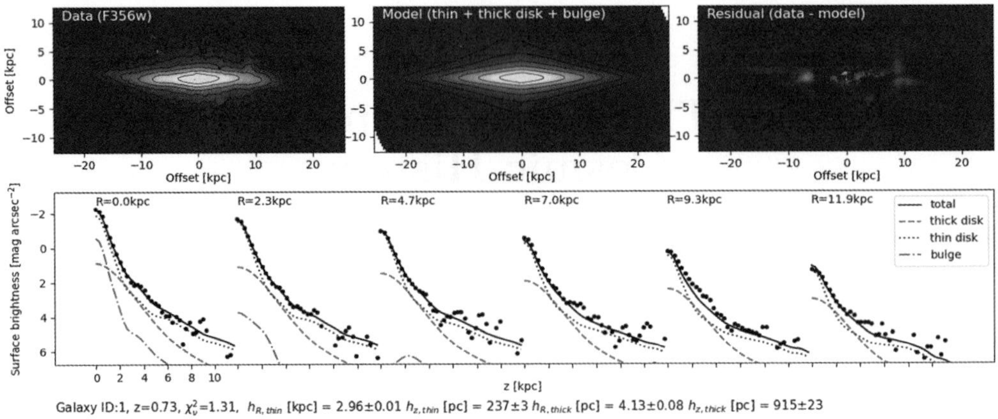

Figure 1. An example fit for a galaxy at $z = 0.73$ with thin and thick disks plus a central bulge. The top row, from left to light shows image, model, and residual, respectively. The bottom row shows the vertical surface brightness profiles at each radius with the decomposed models. This figure is reproduced from Tsukui et al. 2024.

axis-symmetric density distribution, accounting for the JWST point spread function (PSF) and projection on the sky. The edge-on configuration provides maximum signal-to-noise (S/N) to study galaxy surface brightness distributions and enables measurement of intrinsic structural parameters, such as the radial length, h_R, and vertical height, z_0. We use an axial ratio cut of $q < 0.3$ to select edge-on galaxies.

Some galaxies show excess light above the disk mid-plane when modeled with a single disk, suggesting the need for a second disk. Additionally, some galaxies show central excess light from a bulge. We use a set of models such as, single disk, single disk+bulge, two disks, two disks+bulge. Not all galaxies require two disks; in many case, data quality limits the identification of additional components. To classify galaxies objectively as having either one or two disks, we use the Bayesian Information Criterion (BIC), designating galaxies as having two disks if the BIC improves by more than 15. Figure 1 shows an example fit for a galaxy at redshift $z = 0.73$ ($\sim$7 billion years ago) exhibiting both a thin and thick disk.

Several studies have studied vertical disk structures for galaxies at cosmological distances without detecting two distinct components (Elmegreen et al. 2017; Hamilton-Campos et al. 2023; Lian & Luo 2024). The key to detect two disks is the JWST's significant improvement of surface brightness sensitivity and the use of the rest-frame near infrared capability, which is less sensitive to stellar age. Lian & Luo 2024 used the F115W filter (1.2μm), which is sensitive to ongoing star formation (i.e., young stars). In contrast, we use the F277W (2.8μm), F356W (3.6μm), and F444W (4.4μm) filters to approximate rest-frame K or H bands depending on the galaxy redshift. The wavebands trace the stellar mass distribution and provide enhanced contrast to reveal both thick disk and a thin disks.

3. Sequential formation of thick and thin disks

Disks follow a mass-size-height relation already at explored this epoch (Fig. 2). Remarkably, disks in one-disk galaxies have geometric properties h_R/z_0 similar to those of thick disks in two-disk galaxies, suggesting that single disk galaxies host thick disks and then form a thin disk later. This picture is further supported by the distribution of one disk and two disk galaxies in stellar mass and look back time shown in Fig. 4.

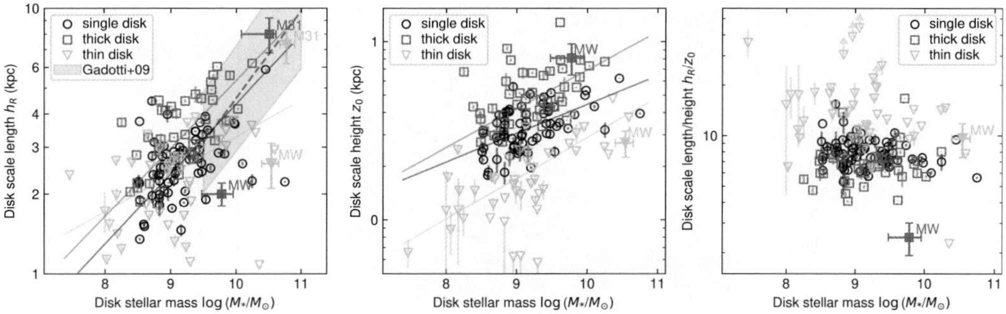

Figure 2. Size-mass relations for disks: single, thin, and thick disks. Left: disk mass versus radial size (h_R); middle: disk mass versus vertical height (z_0); right: disk mass versus the ratio of vertical to radial size (z_0/h_R). This figure is reproduced from Tsukui et al. 2024.

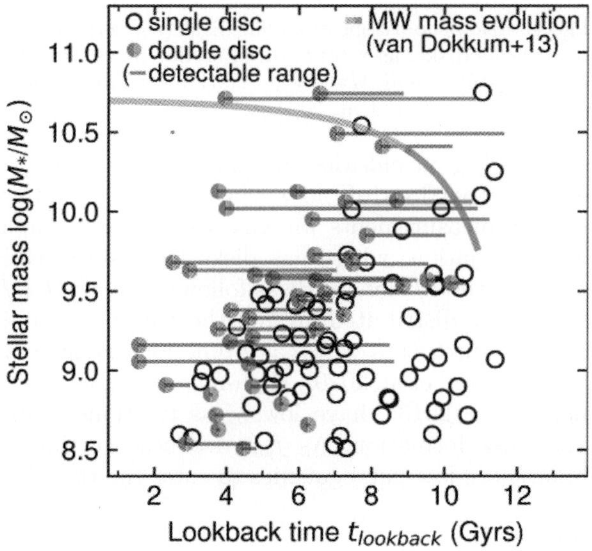

Figure 3. Distribution of galaxies with one-disk and two-disk galaxies in stellar mass and look back time. Estimated redshift (lookback time) range is shown in black lines. This figure is reproduced from Tsukui et al. 2024.

The distribution is partly influenced by observational effects: detecting two disks is easier in high-mass galaxies with higher S/N ratios, while more distant galaxies have lower S/N ratio and thus a reduced likelihood of detecting two disks. To clarify the impact of observational effects, we estimate redshift (lookback time) range we can idenfity two disks, by artificially redshifting two-disk galaxies: surface brightness dimming, galaxy ageing, bandshifting. Guided by the detectable range of two-disks, we find high-mass galaxies formed a thin disk earlier ~ 8 Gyr ago whereas low-mass galaxies formed a thin disk ~ 4 Gyr ago. This aligns with the onset of thin disk formation ~ 9 Gyr ago in the Milky Way (Wu et al. 2023), considering the average mass growth trajectory for the Milky Way-sized galaxies (van Dokkum et al. 2013).

4. Explanation for the downsizing formation of thin disk

The observed downsizing in thin disk formation - high-mass galaxies form thin disks earlier than low-mass galaxies - is a natural consequence of disk stability regulated by

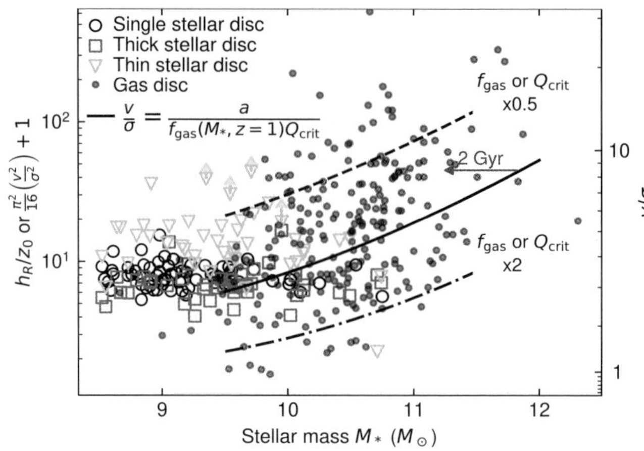

Figure 4. Measured geometric proportions of stellar disks (h_R/z_0) compared with expectations from v/σ measurements of gas disks, as a function of stellar mass $M*$. The expected curve for gas disks is from marginal stability and $M* - f_{\rm gas}$ relation for main-sequence galaxies at $z = 1$. This figure is reproduced from Tsukui et al. 2024.

the evolving gas fraction $f_{\rm gas}$ in galaxies, as illustrated in Fig. 4. This figure shows the measured geometric proportions of our stellar disks, h/z and compares them with those expected from v/σ measurements for gas disks (sources in the footnote †). In disk stability-regulated formation, where gas disks form stars while keeping the disk marginally stable $Q_{\rm crit} \sim 1$, the v/σ of gas disks follows $v/\sigma = a/f_{\rm gas}/Q_{\rm crit}$ (Genzel et al. 2011). In Fig. 4, galaxies are distributed around the marginal stability $Q_{\rm crit} \sim 1$ curve, where we simply adopt measurements of $f_{\rm gas}$ as a function of galaxy mass M_* at $z = 1$ for main sequence galaxies (Tacconi et al. 2020). High-mass galaxies, which efficiently convert gas to stars (Behroozi et al. 2013), have lower gas fractions than low-mass galaxies, supporting earlier thin disk formation. As $f_{\rm gas}$ declines overtime, this stability curve shifts leftward, enabling more low-mass galaxies to enter the thin disk formation regime.

5. Questions asked

Is there any evidence of a bar or pseudo-bulge? Determining the presence of a bar in the disk is challenging from the edge-on orientation. The bulges we see appears to be pseudo-bulge (somewhat due to strict axial ratio cut of $q < 0.3$).

Is it possible that the thick disks observed at $z = 2 - 3$ originally formed as thin disks at $z = 6$ and subsequently heated into thick disks due to internal dynamical evolution? Gas disks with higher gas fractions are expected to be more turbulent forming kinematically hot thick disk (Sec. 4). Given that gas fractions are higher at earlier times, we believe at least the majority of galaxies initially form thick disks.

References

Amvrosiadis, A., Wardlow, J. L., Birkin, J. E., et al. 2023, The kinematics of massive high-redshift dusty star-forming galaxies.

Behroozi, P. S., Wechsler, R. H., & Conroy, C. 2013, *ApJ*, 770, 57.

Belokurov, V., & Kravtsov, A. 2022, *MNRAS*, 514, 689.

Burstein, D. 1979, *ApJ*, 234, 829.

Ciucă, I., Kawata, D., Ting, Y.-S., et al. 2024, *MNRAS*, 528, L122.

† Neeleman et al. 2020; Rizzo et al. 2020, 2021; Lelli et al. 2021; Tsukui & Iguchi 2021; Fraternali et al. 2021; Rizzo et al. 2023; Roman-Oliveira et al. 2023; Parlanti et al. 2023; Amvrosiadis et al. 2023; Rowland et al. 2024; Fujimoto et al. 2024

Conroy, C., Weinberg, D. H., Naidu, R. P., et al. 2022, arXiv e-prints, arXiv:2204.02989.
Elmegreen, B. G., Elmegreen, D. M., Tompkins, B., & Jenks, L. G. 2017, *ApJ*, 847, 14.
Fraternali, F., Karim, A., Magnelli, B., et al. 2021, *A&A*, 647, A194.
Fujimoto, S., Ouchi, M., Kohno, K., et al. 2024, Primordial Rotating Disk Composed of $\geq$15 Dense Star-Forming Clumps at Cosmic Dawn.
Genzel, R., Newman, S., Jones, T., et al. 2011, *ApJ*, 733, 101.
Hamilton-Campos, K. A., Simons, R. C., Peeples, M. S., Snyder, G. F., & Heckman, T. M. 2023, *ApJ*, 956, 147.
Lelli, F., Di Teodoro, E. M., Fraternali, F., et al. 2021, Science, 371, 713.
Lian, J., & Luo, L. 2024, *ApJ*, 960, L10.
Neeleman, M., Prochaska, J. X., Kanekar, N., & Rafelski, M. 2020, *Nature*, 581, 269.
Parlanti, E., Carniani, S., Pallottini, A., et al. 2023, *A&A*, 673, A153.
Rizzo, F., Vegetti, S., Fraternali, F., Stacey, H. R., & Powell, D. 2021, *MNRAS*, 507, 3952.
Rizzo, F., Vegetti, S., Powell, D., et al. 2020, *Nature*, 584, 201.
Rizzo, F., Roman-Oliveira, F., Fraternali, F., et al. 2023, *A&A*, 679, A129.
Roman-Oliveira, F., Fraternali, F., & Rizzo, F. 2023, *MNRAS*, 521, 1045.
Rowland, L. E., Hodge, J., Bouwens, R., et al. 2024, REBELS-25: Discovery of a dynamically cold disc galaxy at z = 7.31, arXiv. http://arxiv.org/abs/2405.06025
Tacconi, L. J., Genzel, R., & Sternberg, A. 2020, *ARA&A*, 58, 157.
Tsukui, T., & Iguchi, S. 2021, Science, 372, 1201.
Tsukui, T., Wisnioski, E., Bland-Hawthorn, J., & Freeman, K. 2024, arXiv e-prints, arXiv:2409.15909.
van Dokkum, P. G., Leja, J., Nelson, E. J., et al. 2013, *ApJ*, 771, L35.
Vincenzo, F., Spitoni, E., Calura, F., et al. 2019, *MNRAS*, 487, L47.
Wu, Y., Xiang, M., Zhao, G., et al. 2023, *MNRAS*, 520, 1913.
Yoachim, P., & Dalcanton, J. J. 2006, *AJ*, 131, 226.

The First Chapters of Our Cosmic History with JWST
Proceedings IAU Symposium No. 391, 2025
T. Nanayakkara & M. Maseda, eds.
doi:10.1017/S1743921324002357

JWST/NIRSpec IFU observations show evidence of metal-rich, massive galaxies in the Epoch of Reionisation

Lucie E. Rowland

Leiden Observatory, Leiden University, P.O. Box 9513, 2300 RA Leiden, The Netherlands
email: lrowland@strw.leidenuniv.nl

Abstract. The REBELS ALMA large program has obtained, to date, the largest number of [C II] and dust continuum observations of massive galaxies at z 6-8. We present results from a now complete JWST NIRSpec/IFU program (GO-1626, PI Stefanon) targeting 12 of these galaxies at 0.6-5.3μm. Due to the incredible sensitivity of the NIRSpec prism, key optical emission lines between [O II] and [S II] are detected for the majority of the sample, enabling metallicity constraints from a range of different calibrations. One surprising finding of this dataset is the near-solar oxygen abundances for several of these galaxies. We focus on an extension of the mass-metallicity relations at high-z to higher stellar masses, in comparison to the typical galaxies observed in e.g., JADES and CEERS. These relations provide fundamental information on how stellar mass is assembled in galaxies, and how star formation enriches and dilutes the interstellar medium.

Keywords. galaxies: high-redshift, galaxies: ISM, galaxies: star formation

1. Introduction

A growing number of star-forming galaxies have now been identified at redshifts $z > 10$ (e.g. Oesch et al. 2016; Jiang et al. 2021; Curti et al. 2023; Robertson et al. 2023; Bunker et al. 2023; Carniani et al. 2024), and at slightly later cosmic times ($z \sim 6$-8, in the EoR) there have been several discoveries of massive galaxies ($M_* \sim 10^9$-$10^{11}\mathrm{M}_\odot$, e.g. Strandet et al. 2017; Marrone et al. 2018; Bañados et al. 2018; Bouwens et al. 2022) and substantial dust reservoirs (e.g. Tamura et al. 2019; Schouws et al. 2022; Inami et al. 2022). This raises fundamental questions on how such large systems could have formed so quickly.

The Reionisation Era Bright Emission Line Survey (REBELS, Bouwens et al. 2022) aims to help address these questions about the build-up of massive galaxies in the early Universe. This is a now-complete cycle-7 ALMA large program that targeted $\sim$ 40 UV-bright galaxies with masses $\log(M_*/\mathrm{M}_\odot) \sim 9 - 10$ by carrying out spectral scans in ALMA's band 6 for the [C II]157.7μm emission line. The program resulted in 16 galaxies with confirmed $\sim 150\mu$m continuum emissions and 25 galaxies with [C II] detections (Schouws et al. in prep, Inami et al. 2022), spectroscopically confirming their redshifts at $z = 6.5 - 8$ and providing insights into the dust and neutral gas content of star forming galaxies in the EoR (Ferrara et al. 2022; Dayal et al. 2013; Sommovigo et al. 2022; Fudamoto et al. 2022). However, prior to the launch of *JWST*, constraining the ionised gas properties of galaxies at such high redshifts remained challenging. In particular, the gas-phase metallicity of galaxies, as traced by the oxygen abundance, provides a crucial window into their evolutionary history, but typically requires the detection of rest-frame optical emission lines that are absorbed by the Earth's atmosphere at these redshifts.

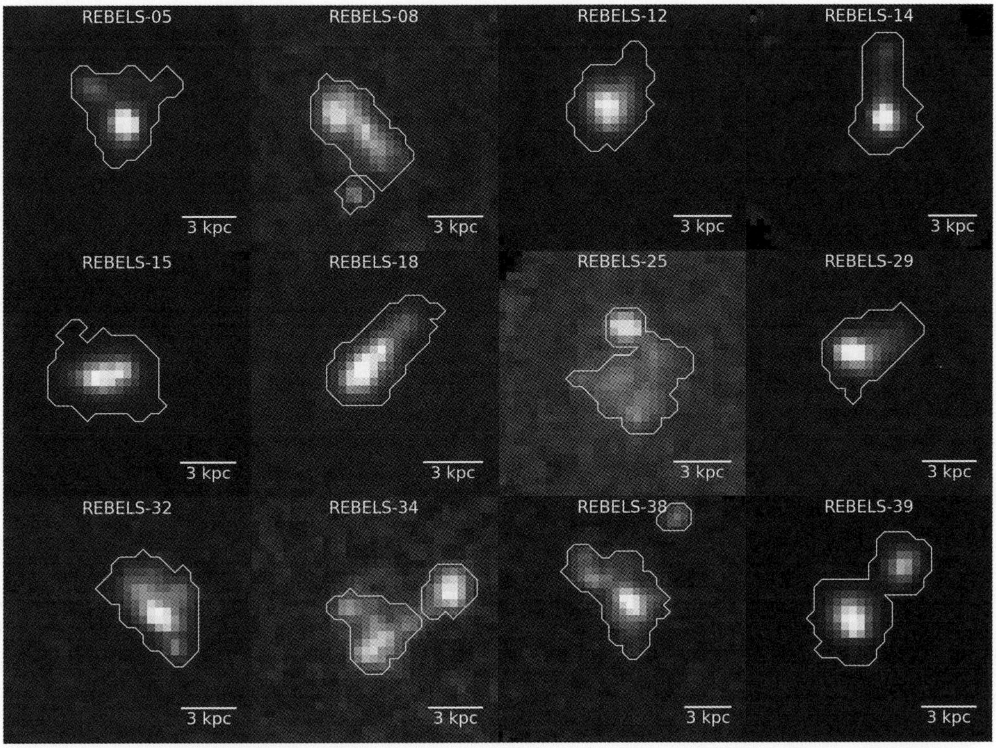

Figure 1. Three-colour composite images of each of the 12 REBELS galaxies targeted in the Cycle 1 *JWST* program presented in this work. Red corresponds to rest-frame optical ($0.37 < \lambda_{\rm rest} < 0.855\mu$m), green to [O III]$\lambda$5007 and blue to rest-frame UV emission ($0.12 < \lambda_{\rm rest} < 0.37\mu$m). The white contours are used to extract the integrated spectrum of each galaxy.

2. *JWST*'s view of the REBELS galaxies

The launch of *JWST* has now enabled the routine detection of rest-frame optical emission lines at $z > 6$ (e.g. Arellano-Córdova et al. 2022; Curti et al. 2022; Katz et al. 2023b; Taylor et al. 2022; Trump et al. 2022; Schaerer et al. 2022; Brinchmann 2023; Carnall et al. 2023), including for 12 of the REBELS galaxies. We present some of the first results from a *JWST* Cycle 1 program (GO-1626, PI M. Stefanon) targeting 11 [C II]-bright ($L_{\rm [C\,II]} \gtrsim 2 \times 10^8 {\rm L}_\odot$) REBELS galaxies at redshifts $z = 6.496$ to 7.349, with an additional REBELS galaxy from GO 2659 PI J. Weaver at $z = 7.675$. The observations were carried out with the NIRSpec/IFU instrument in the prism mode, covering an observed wavelength range of 0.6 to 5.3μm with a spectral resolution of $\mathcal{R} \sim 100$. Thanks to a combination of NIRSpec's broad wavelength range, exceptional sensitivity, and the intrinsic brightness of these sources, a wealth of key rest-frame optical emission lines from [O II]λ3727, 9 to [S II]λ6716, 33 were detected for the majority of the sample, enabling properties like their oxygen abundance to be constrained for the first time. An overview of this program was presented in another talk in this Symposium (M. Stefanon: Divide et Impera: Galaxy Assembly in the Early Universe from NIRSpec/IFU Observations of ALMA/REBELS Galaxies), which we refer to for more details.

We show in Figure 1 three-colour composite images of each of the galaxies targeted in this program, with red showing rest-frame optical emission ($0.37 < \lambda_{\rm rest} < 0.855\mu$m), green [O III]$\lambda$5007 emission, and in blue the rest frame UV emission ($0.12 < \lambda_{\rm rest} < 0.37\mu$m). This powerful synergy of *JWST* and ALMA means that we have both spectral and spatial coverage of the cold neutral gas, dust content, ionised gas, stellar populations,

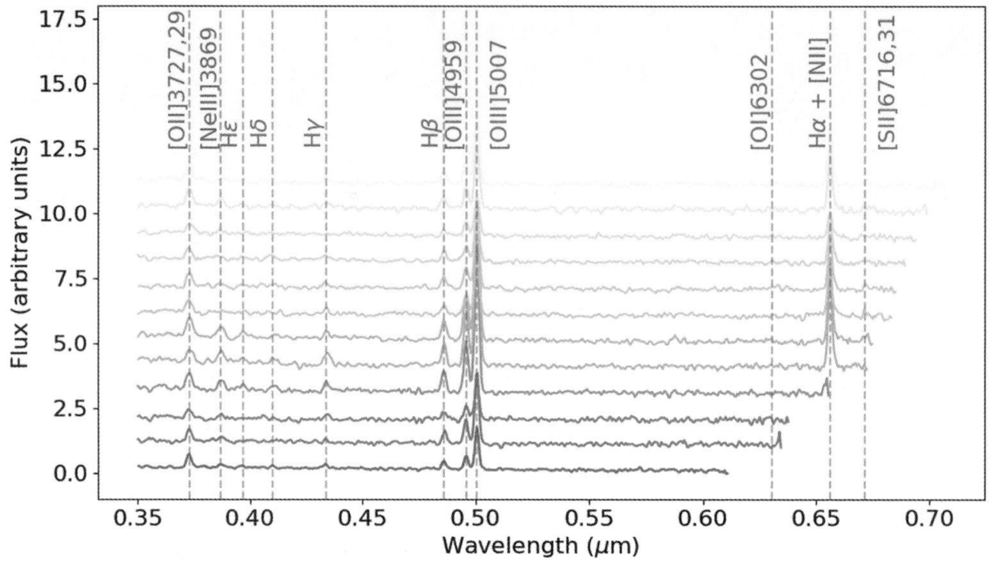

Figure 2. The integrated NIRSpec spectrum of each of the REBELS galaxies, extracted from a target-specific aperture mask shown in Figure 1. Each spectrum is cropped to $\lambda_{\rm rest} = 0.35\mu$m to $\lambda_{\rm obs}$ to highlight the key detected emission lines, labelled via the grey dashed lines and text. The fluxes are normalised to the Hβ flux of each galaxy, and shifted in the y-axis for visual clarity. Lower redshift galaxies are plotted in yellow, and the higher redshift galaxies in red.

and metal content within these 12 EoR galaxies. We see a variety of morphologies, both across the sample and at different wavelengths for individual galaxies. In particular, whilst some galaxies look clumpy in the rest-frame UV, high resolution [C II] observations with ALMA reveal more disc-like morphologies, with more compact dust emission (Rowland et al. 2024).

These detailed observations allow us not only to map out the physical structures of these galaxies, but also to probe their chemical enrichment. In Figure 2, we show the integrated spectrum of each galaxy at $\lambda_{\rm rest} > 0.35\mu$m to highlight the detected emission lines which we use to derive the oxygen abundances.

3. Evidence for metal-rich galaxies in the EoR

With key emission lines like [O II]λ3727, 9, Hβ, and [O III]λ4959, 5007 detected for all 12 of the REBELS galaxies, we are able to use the ratios of these lines (predominantly R23 and R3, but also O32, Ne3O2, N2 and S2) with strong-line calibrations in the literature (Maiolino et al. 2008; Bian et al. 2021; Curti et al. 2020; Nakajima et al. 2022; Sanders et al. 2024) to derive the oxygen abundances of these sources. As our fiducial estimates, we adopt the R23 and R3 indices with the Sanders et al. (2024) calibrations. We find that the REBELS galaxies are more metal-rich than the general population at $z \sim 6$-8, with some galaxies potentially exhibiting near- or even super-solar abundances just ~ 700 Myr after the Big Bang. However, it is important to note that these strong-line calibrations come with significant uncertainties. More auroral line detections, which enable direct measurements of electron temperature and, consequently, metallicity, will be essential for refining these calibrations and confirming these high metallicities.

In Figure 3, we show the oxygen abundances as a function of stellar mass for the REBELS sample and for a sample of $z \sim 6$-8 galaxies from the literature (Nakajima et al. 2023; Chemerynska et al. 2024) where we have re-derived their metallicities using the same methodology for consistency. The full details of our data reduction, emission line fitting, SED fitting, and derivation of ISM properties will be detailed in subsequent

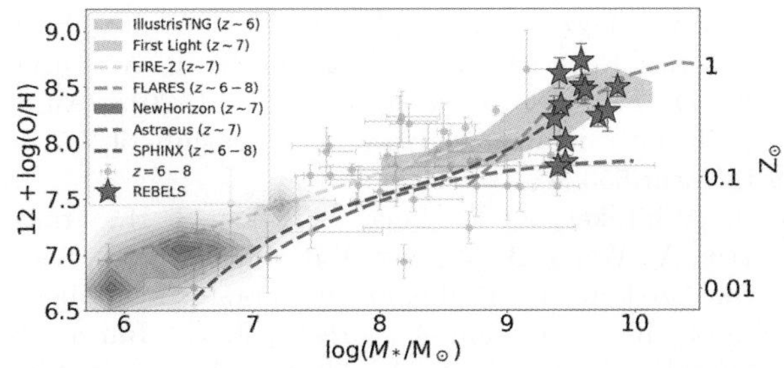

Figure 3. Stellar mass-gas-phase oxygen abundance relation (MZR) at $z \sim$ 6-8 for the REBELS galaxies and other literature sources derived in Nakajima et al. (2023) from CEERS (Finkelstein et al. 2023) GLASS (Roberts-Borsani et al. 2024), and ERO (Pontoppidan et al. 2022) observations, and in Chemerynska et al. 2024 from UNCOVER (Bezanson et al. 2024). We also show the predicted MZR at $z \sim$ 6-8 from the following cosmological simulations: NEWHORIZON (Dubois et al. 2021), IllustrisTNG (Torrey et al. 2019), First Light (Nakazato et al. 2023), FIRE-2 (Marszewski et al. 2024), FLARES (Wilkins et al. 2023), Astraeus (Ucci et al. 2023), and SPHINX (Katz et al. 2023a).

works (Stefanon et al. in prep, Rowland et al. in prep). The correlation between the metallicity and stellar mass of galaxies, the so-called mass-metallicity relation (MZR), is a well-established scaling relation that serves as a key diagnostic of galaxy evolution, and in particular of processes like gas accretion, star formation, and feedback-driven inflows/outflows (e.g., Peeples and Shankar (2011); Lilly et al. (2013); Torrey et al. (2019)). It was first observed in the local Universe (e.g. McClure and van den Bergh 1968; Trager et al. 2000, also see review by Maiolino and Mannucci 2019 and references therein), but has now been observed out as far as $z \sim 10$ (e.g. Nakajima et al. 2023; Heintz et al. 2023; Curti et al. 2023; Chemerynska et al. 2024; Sarkar et al. 2024). However, these high-z studies are typically limited to stellar masses $\log(M_*/\mathrm{M}_\odot) \lesssim 9$. With the REBELS galaxies probing a higher mass range, we can now constrain the MZR over a ~ 4 dex range in stellar mass at $z \sim$ 6-8 and compare to predictions from cosmological simulations, which we also show in Figure 3.

4. Summary & Outlook

We have presented a first-look at the UV-to-optical morphology and integrated spectra of 12 massive, [C II]-luminous EoR galaxies selected from the REBELS ALMA large program. With the detected rest-frame optical emission lines, we infer near-solar oxygen abundances for some of these galaxies, indicating a population of evolved, metal-rich galaxies in the first Gyr of cosmic history (Rowland et al. in prep; Stefanon et al. in prep). In subsequent works, we will utilise these *JWST* observations, in combination with ALMA observations of the same galaxies, to investigate the build-up of massive galaxies in the early Universe. In particular, we will study these galaxies in the context of the redshift evolution of the MZR and other metallicity scaling relations (Rowland et al. in prep), and derive their dust-to-gas, dust-to-metal, and dust-to-stellar mass ratios in order to investigate the dust production processes at these early cosmic times (Algera et al. in prep).

References

Arellano-Córdova, K. Z., Berg, D. A., Chisholm, J., Haro, P. A., Dickinson, M., Finkelstein, S. L., Leclercq, F., Rogers, N. S. J., Simons, R. C., Skillman, E. D., Trump, J. R., & Kartaltepe, J. S. 2022,. A First Look at the Abundance Pattern – O/H, C/O, and Ne/O – in $z > 7$ Galaxies with JWST/NIRSpec. arXiv:2208.02562.

Bañados, E., Venemans, B. P., Mazzucchelli, C., Farina, E. P., Walter, F., Wang, F., Decarli, R., Stern, D., Fan, X., Davies, F. B., Hennawi, J. F., Simcoe, R. A., Turner, M. L., Rix, H.-W., Yang, J., Kelson, D. D., Rudie, G. C., & Winters, J. M. 2018, An 800-million-solar-mass black hole in a significantly neutral Universe at a redshift of 7.5. *Nature*, 553, 473–476. ADS Bibcode: 2018Natur.553..473B.

Bezanson, R., Labbe, I., Whitaker, K. E., Leja, J., Price, S. H., Franx, M., Brammer, G., Marchesini, D., Zitrin, A., Wang, B., Weaver, J. R., Furtak, L. J., Atek, H., Coe, D., Cutler, S. E., Dayal, P., van Dokkum, P., Feldmann, R., Förster Schreiber, N. M., Fujimoto, S., Geha, M., Glazebrook, K., de Graaff, A., Greene, J. E., Juneau, S., Kassin, S., Kriek, M., Khullar, G., Maseda, M., Mowla, L. A., Muzzin, A., Nanayakkara, T., Nelson, E. J., Oesch, P. A., Pacifici, C., Pan, R., Papovich, C., Setton, D. J., Shapley, A. E., Smit, R., Stefanon, M., Taylor, E. N., & Williams, C. C. 2024, The JWST UNCOVER Treasury Survey: Ultradeep NIRSpec and NIRCam Observations before the Epoch of Reionization. *The Astrophysical Journal*, 974, 92. Publisher: IOP ADS Bibcode: 2024ApJ...974...92B.

Bian, F., Kewley, L. J., & Dopita, M. A. 2021, "Direct" Gas-Phase Metallicity in Local Analogs of High-Redshift Galaxies: Empirical Metallicity Calibrations for High-Redshift Star-Forming Galaxies. arXiv:1805.08224.

Bouwens, R. J., Smit, R., Schouws, S., Stefanon, M., Bowler, R., Endsley, R., Gonzalez, V., Inami, H., Stark, D., Oesch, P., Hodge, J., Aravena, M., da Cunha, E., Dayal, P., de Looze, I., Ferrara, A., Fudamoto, Y., Graziani, L., Li, C., Nanayakkara, T., Pallotini, A., Schneider, R., Sommovigo, L., Topping, M., van der Werf, P., Algera, H., Barrufet, L., Hygate, A., Labbe, I., Riechers, D., & Witstok, J. 2022, Reionization Era Bright Emission Line Survey: Selection and Characterization of Luminous Interstellar Medium Reservoirs in the z>6.5 Universe. *ApJ*, 931(2), 160. arXiv:2106.13719 [astro-ph].

Brinchmann, J. 2023, High-z galaxies with JWST and local analogues - it is not only star formation. *Monthly Notices of the Royal Astronomical Society*, 525(2), 2087–2106.

Bunker, A. J., Saxena, A., Cameron, A. J., Willott, C. J., Curtis-Lake, E., Jakobsen, P., Carniani, S., Smit, R., Maiolino, R., Witstok, J., Curti, M., D'Eugenio, F., Jones, G. C., Ferruit, P., Arribas, S., Charlot, S., Chevallard, J., Giardino, G., de Graaff, A., Looser, T. J., Lützgendorf, N., Maseda, M. V., Rawle, T., Rix, H.-W., Del Pino, B. R., Alberts, S., Egami, E., Eisenstein, D. J., Endsley, R., Hainline, K., Hausen, R., Johnson, B. D., Rieke, G., Rieke, M., Robertson, B. E., Shivaei, I., Stark, D. P., Sun, F., Tacchella, S., Tang, M., Williams, C. C., Willmer, C. N. A., Baker, W. M., Baum, S., Bhatawdekar, R., Bowler, R., Boyett, K., Chen, Z., Circosta, C., Helton, J. M., Ji, Z., Kumari, N., Lyu, J., Nelson, E., Parlanti, E., Perna, M., Sandles, L., Scholtz, J., Suess, K. A., Topping, M. W., Übler, H., Wallace, I. E. B., & Whitler, L. 2023, JADES NIRSpec Spectroscopy of GN-z11: Lyman-α emission and possible enhanced nitrogen abundance in a z = 10.60 luminous galaxy. *Astronomy and Astrophysics*, 677, A88. ADS Bibcode: 2023A&A...677A..88B.

Carnall, A. C., Begley, R., McLeod, D. J., Hamadouche, M. L., Donnan, C. T., McLure, R. J., Dunlop, J. S., Milvang-Jensen, B., Bondestam, C. L., Cullen, F., Jewell, S. M., & Pollock, C. L. 2023, A first look at the SMACS0723 JWST ERO: spectroscopic redshifts, stellar masses, and star-formation histories. *Monthly Notices of the Royal Astronomical Society*, 518, L45–L50. Publisher: OUP ADS Bibcode: 2023MNRAS.518L..45C.

Carniani, S., Hainline, K., D'Eugenio, F., Eisenstein, D. J., Jakobsen, P., Witstok, J., Johnson, B. D., Chevallard, J., Maiolino, R., Helton, J. M., Willott, C., Robertson, B., Alberts, S., Arribas, S., Baker, W. M., Bhatawdekar, R., Boyett, K., Bunker, A. J., Cameron, A. J., Cargile, P. A., Charlot, S., Curti, M., Curtis-Lake, E., Egami, E., Giardino, G., Isaak, K., Ji, Z., Jones, G. C., Kumari, N., Maseda, M. V., Parlanti, E., Pérez-González, P. G., Rawle, T., Rieke, G., Rieke, M., Del Pino, B. R., Saxena, A., Scholtz, J., Smit, R., Sun, F., Tacchella, S., Übler, H., Venturi, G., Williams, C. C., & Willmer, C. N. A. 2024, Spectroscopic confirmation of two luminous galaxies at a redshift of 14. *Nature*, 633, 318–322. ADS Bibcode: 2024Natur.633..318C.

Chemerynska, I., Atek, H., Dayal, P., Furtak, L. J., Feldmann, R., Greene, J. E., Maseda, M. V., Nanayakkara, T., Oesch, P. A., Labbe, I., Bezanson, R., Brammer, G., Cutler, S. E.,

Leja, J., Pan, R., Price, S. H., Wang, B., Weaver, J. R., & Whitaker, K. E. 2024,. The Extreme Low-mass End of the Mass-Metallicity Relation at $z\sim7$. arXiv:2407.17110 [astro-ph].

Curti, M., D'Eugenio, F., Carniani, S., Maiolino, R., Sandles, L., Witstok, J., Baker, W. M., Bennett, J. S., Piotrowska, J. M., Tacchella, S., Charlot, S., Nakajima, K., Maheson, G., Mannucci, F., Amiri, A., Arribas, S., Belfiore, F., Bonaventura, N. R., Bunker, A. J., Chevallard, J., Cresci, G., Curtis-Lake, E., Hayden-Pawson, C., Kumari, N., Laseter, I., Looser, T. J., Marconi, A., Maseda, M. V., Jones, G. C., Scholtz, J., Smit, R., Ubler, H., & Wallace, I. E. B. 2022, The chemical enrichment in the early Universe as probed by JWST via direct metallicity measurements at z˜8. *Monthly Notices of the Royal Astronomical Society*, 518(1), 425–438. arXiv:2207.12375 [astro-ph].

Curti, M., Maiolino, R., Curtis-Lake, E., Chevallard, J., Carniani, S., D'Eugenio, F., Looser, T. J., Scholtz, J., Charlot, S., Cameron, A., Übler, H., Witstok, J., Boyett, K., Laseter, I., Sandles, L., Arribas, S., Bunker, A., Giardino, G., Maseda, M. V., Rawle, T., Del Pino, B. R., Smit, R., Willott, C. J., Eisenstein, D. J., Hausen, R., Johnson, B., Rieke, M., Robertson, B., Tacchella, S., Williams, C. C., Willmer, C., Baker, W. M., Bhatawdekar, R., Egami, E., Helton, J. M., Ji, Z., Kumari, N., Perna, M., Shivaei, I., & Sun, F. 2023,. JADES: Insights on the low-mass end of the mass–metallicity–star-formation rate relation at $3 < z < 10$ from deep JWST/NIRSpec spectroscopy. arXiv:2304.08516 [astro-ph].

Curti, M., Mannucci, F., Cresci, G., & Maiolino, R. 2020, The mass-metallicity and the fundamental metallicity relation revisited on a fully Te-based abundance scale for galaxies. *Monthly Notices of the Royal Astronomical Society*, 491, 944–964. Publisher: OUP ADS Bibcode: 2020MNRAS.491..944C.

Dayal, P., Ferrara, A., & Dunlop, J. S. 2013, The physics of the fundamental metallicity relation. *Monthly Notices of the Royal Astronomical Society*, 430(4), 2891–2895. arXiv:1202.4770 [astro-ph].

Dubois, Y., Beckmann, R., Bournaud, F., Choi, H., Devriendt, J., Jackson, R., Kaviraj, S., Kimm, T., Kraljic, K., Laigle, C., Martin, G., Park, M.-J., Peirani, S., Pichon, C., Volonteri, M., & Yi, S. K. 2021, Introducing the NEWHORIZON simulation: Galaxy properties with resolved internal dynamics across cosmic time. *Astronomy and Astrophysics*, 651, A109. ADS Bibcode: 2021A&A...651A.109D.

Ferrara, A., Sommovigo, L., Dayal, P., Pallottini, A., Bouwens, R. J., Gonzalez, V., Inami, H., Smit, R., Bowler, R. A. A., Endsley, R., Oesch, P., Schouws, S., Stark, D., Stefanon, M., Aravena, M., da Cunha, E., De Looze, I., Fudamoto, Y., Graziani, L., Hodge, J., Riechers, D., Schneider, R., Algera, H. S. B., Barrufet, L., Hygate, A. P. S., Labbé, I., Li, C., Nanayakkara, T., Topping, M., & van der Werf, P. 2022, The ALMA REBELS Survey. Epoch of Reionization giants: Properties of dusty galaxies at z $\approx$ 7. *Monthly Notices of the Royal Astronomical Society*, 512, 58–72. Publisher: OUP ADS Bibcode: 2022MNRAS.512...58F.

Finkelstein, S. L., Bagley, M. B., Ferguson, H. C., Wilkins, S. M., Kartaltepe, J. S., Papovich, C., Yung, L. Y. A., Haro, P. A., Behroozi, P., Dickinson, M., Kocevski, D. D., Koekemoer, A. M., Larson, R. L., Bail, A. L., Morales, A. M., Perez-Gonzalez, P. G., Burgarella, D., Dave, R., Hirschmann, M., Somerville, R. S., Wuyts, S., Bromm, V., Casey, C. M., Fontana, A., Fujimoto, S., Gardner, J. P., Giavalisco, M., Grazian, A., Grogin, N. A., Hathi, N. P., Hutchison, T. A., Jha, S. W., Jogee, S., Kewley, L. J., Kirkpatrick, A., Long, A. S., Lotz, J. M., Pentericci, L., Pierel, J. D. R., Pirzkal, N., Ravindranath, S., Ryan Jr, R. E., Trump, J. R., Yang, G., Bhatawdekar, R., Bisigello, L., Buat, V., Calabro, A., Castellano, M., Cleri, N. J., Cooper, M. C., Croton, D., Daddi, E., Dekel, A., Elbaz, D., Franco, M., Gawiser, E., Holwerda, B. W., Huertas-Company, M., Jaskot, A. E., Leung, G. C. K., Lucas, R. A., Mobasher, B., Pandya, V., Tacchella, S., Weiner, B. J., & Zavala, J. A. 2023, CEERS Key Paper I: An Early Look into the First 500 Myr of Galaxy Formation with JWST. *ApJL*, 946(1), L13. arXiv:2211.05792 [astro-ph].

Fudamoto, Y., Smit, R., Bowler, R. A. A., Oesch, P. A., Bouwens, R., Stefanon, M., Inami, H., Endsley, R., Gonzalez, V., Schouws, S., Stark, D., Algera, H. S. B., Aravena, M., Barrufet, L., da Cunha, E., Dayal, P., Ferrara, A., Graziani, L., Hodge, J. A., Hygate, A. P. S., Inoue, A. K., Nanayakkara, T., Pallottini, A., Pizzati, E., Schneider, R., Sommovigo, L., Sugahara, Y., Topping, M., van der Werf, P., Bethermin, M., Cassata, P., Dessauges-Zavadsky, M., Ibar, E., Faisst, A. L., Fujimoto, S., Ginolfi, M., Hathi, N., Jones, G. C., Pozzi, F., & Schaerer, D. 2022, The ALMA REBELS Survey: Average [C II] 158μm Sizes of Star-forming Galaxies from z 7 to z 4. *The Astrophysical Journal*, 934, 144. Publisher: IOP ADS Bibcode: 2022ApJ...934..144F.

Heintz, K. E., Brammer, G. B., Giménez-Arteaga, C., Strait, V. B., del P. Lagos, C., Vijayan, A. P., Matthee, J., Watson, D., Mason, C. A., Hutter, A., Toft, S., Fynbo, J. P. U., & Oesch, P. A. 2023, Dilution of chemical enrichment in galaxies 600 Myr after the Big Bang. *Nature Astronomy*, 7, 1517–1524. ADS Bibcode: 2023NatAs...7.1517H.

Inami, H., Algera, H. S. B., Schouws, S., Sommovigo, L., Bouwens, R., Smit, R., Stefanon, M., Bowler, R. A. A., Endsley, R., Ferrara, A., Oesch, P., Stark, D., Aravena, M., Barrufet, L., da Cunha, E., Dayal, P., De Looze, I., Fudamoto, Y., Gonzalez, V., Graziani, L., Hodge, J. A., Hygate, A. P. S., Nanayakkara, T., Pallottini, A., Riechers, D. A., Schneider, R., Topping, M., & van der Werf, P. 2022, The ALMA REBELS Survey: Dust Continuum Detections at $z > 6.5$. *Monthly Notices of the Royal Astronomical Society*, 515(3), 3126–3143. arXiv:2203.15136 [astro-ph].

Jiang, L., Kashikawa, N., Wang, S., Walth, G., Ho, L. C., Cai, Z., Egami, E., Fan, X., Ito, K., Liang, Y., Schaerer, D., & Stark, D. P. 2021, Evidence for GN-z11 as a luminous galaxy at redshift 10.957. *Nature Astronomy*, 5, 256–261. ADS Bibcode: 2021NatAs...5..256J.

Katz, H., Rosdahl, J., Kimm, T., Blaizot, J., Choustikov, N., Farcy, M., Garel, T., Haehnelt, M. G., Michel-Dansac, L., & Ocvirk, P. 2023,a The SPHINX Public Data Release: Forward Modelling High-Redshift JWST Observations with Cosmological Radiation Hydrodynamics Simulations. *The Open Journal of Astrophysics*, 6a. Publisher: Maynooth Academic Publishing.

Katz, H., Saxena, A., Cameron, A. J., Carniani, S., Bunker, A. J., Arribas, S., Bhatawdekar, R., Bowler, R. A. A., Boyett, K. N. K., Cresci, G., Curtis-Lake, E., D'Eugenio, F., Kumari, N., Looser, T. J., Maiolino, R., Übler, H., Willott, C., & Witstok, J. 2023,b First insights into the ISM at $z > 8$ with JWST: possible physical implications of a high [O III] λ4363/[O III] λ5007. *Monthly Notices of the Royal Astronomical Society*, 518b, 592–603. Publisher: OUP ADS Bibcode: 2023MNRAS.518..592K.

Lilly, S. J., Carollo, C. M., Pipino, A., Renzini, A., & Peng, Y. 2013, Gas Regulation of Galaxies: The Evolution of the Cosmic Specific Star Formation Rate, the Metallicity-Mass-Star-formation Rate Relation, and the Stellar Content of Halos. *The Astrophysical Journal*, 772, 119. Publisher: IOP ADS Bibcode: 2013ApJ...772..119L.

Maiolino, R. & Mannucci, F. 2019, De Re Metallica: The cosmic chemical evolution of galaxies. *Astron Astrophys Rev*, 27(1), 3. arXiv:1811.09642 [astro-ph].

Maiolino, R., Nagao, T., Grazian, A., Cocchia, F., Marconi, A., Mannucci, F., Cimatti, A., Pipino, A., Ballero, S., Calura, F., Chiappini, C., Fontana, A., Granato, G. L., Matteucci, F., Pastorini, G., Pentericci, L., Risaliti, G., Salvati, M., & Silva, L. 2008,. AMAZE. I. The evolution of the mass-metallicity relation at $z > 3$. arXiv:0806.2410.

Marrone, D. P., Spilker, J. S., Hayward, C. C., Vieira, J. D., Aravena, M., Ashby, M. L. N., Bayliss, M. B., Béthermin, M., Brodwin, M., Bothwell, M. S., Carlstrom, J. E., Chapman, S. C., Chen, C.-C., Crawford, T. M., Cunningham, D. J. M., De Breuck, C., Fassnacht, C. D., Gonzalez, A. H., Greve, T. R., Hezaveh, Y. D., Lacaille, K., Litke, K. C., Lower, S., Ma, J., Malkan, M., Miller, T. B., Morningstar, W. R., Murphy, E. J., Narayanan, D., Phadke, K. A., Rotermund, K. M., Sreevani, J., Stalder, B., Stark, A. A., Strandet, M. L., Tang, M., & Weiß, A. 2018, Galaxy growth in a massive halo in the first billion years of cosmic history. *Nature*, 553, 51–54. ADS Bibcode: 2018Natur.553...51M.

Marszewski, A., Sun, G., Faucher-Giguère, C.-A., Hayward, C. C., & Feldmann, R. 2024,. The High-Redshift Gas-Phase Mass-Metallicity Relation in FIRE-2. arXiv:2403.08853 [astro-ph].

McClure, R. D. & van den Bergh, S. 1968, UB V Observations of Field Galaxies. *The Astronomical Journal*, 73, 1008. Publisher: IOP ADS Bibcode: 1968AJ.....73.1008M.

Nakajima, K., Ouchi, M., Isobe, Y., Harikane, Y., Zhang, Y., Ono, Y., Umeda, H., & Oguri, M. 2023,. JWST Census for the Mass-Metallicity Star-Formation Relations at z=4-10 with the Self-Consistent Flux Calibration and the Proper Metallicity Calibrators. arXiv:2301.12825 [astro-ph].

Nakajima, K., Ouchi, M., Xu, Y., Rauch, M., Harikane, Y., Nishigaki, M., Isobe, Y., Kusakabe, H., Nagao, T., Ono, Y., Onodera, M., Sugahara, Y., Kim, J. H., Komiyama, Y., Lee, C.-H., & Zahedy, F. S. 2022, EMPRESS. V. Metallicity Diagnostics of Galaxies over 12+log(O/H)=∼6.9-8.9 Established by a Local Galaxy Census: Preparing for JWST Spectroscopy. *ApJS*, 262(1), 3. arXiv:2206.02824 [astro-ph].

Nakazato, Y., Yoshida, N., & Ceverino, D. 2023, Simulations of High-redshift [O III] Emitters: Chemical Evolution and Multiline Diagnostics. *The Astrophysical Journal*, 953, 140. Publisher: IOP ADS Bibcode: 2023ApJ...953..140N.

Oesch, P. A., Brammer, G., van Dokkum, P. G., Illingworth, G. D., Bouwens, R. J., Labbé, I., Franx, M., Momcheva, I., Ashby, M. L. N., Fazio, G. G., Gonzalez, V., Holden, B., Magee, D., Skelton, R. E., Smit, R., Spitler, L. R., Trenti, M., & Willner, S. P. 2016, A Remarkably Luminous Galaxy at z=11.1 Measured with Hubble Space Telescope Grism Spectroscopy. *The Astrophysical Journal*, 819, 129. Publisher: IOP ADS Bibcode: 2016ApJ...819..129O.

Peeples, M. S. & Shankar, F. 2011, Constraints on star-formation driven galaxy winds from the mass-metallicity relation at z=0. *Monthly Notices of the Royal Astronomical Society*, 417(4), 2962–2981. arXiv:1007.3743 [astro-ph].

Pontoppidan, K. M., Barrientes, J., Blome, C., Braun, H., Brown, M., Carruthers, M., Coe, D., DePasquale, J., Espinoza, N., Marin, M. G., Gordon, K. D., Henry, A., Hustak, L., James, A., Jenkins, A., Koekemoer, A. M., LaMassa, S., Law, D., Lockwood, A., Moro-Martin, A., Mullally, S. E., Pagan, A., Player, D., Proffitt, C., Pulliam, C., Ramsay, L., Ravindranath, S., Reid, N., Robberto, M., Sabbi, E., Ubeda, L., Balogh, M., Flanagan, K., Gardner, J., Hasan, H., Meinke, B., & Nota, A. 2022, The JWST Early Release Observations. *The Astrophysical Journal*, 936, L14. Publisher: IOP ADS Bibcode: 2022ApJ...936L..14P.

Roberts-Borsani, G., Treu, T., Shapley, A., Fontana, A., Pentericci, L., Castellano, M., Morishita, T., Bergamini, P., & Rosati, P. 2024,. Between the Extremes: A JWST Spectroscopic Benchmark for High Redshift Galaxies Using ˜500 Confirmed Sources at \$z\geqslant5\$. Publication Title: arXiv e-prints ADS Bibcode: 2024arXiv240307103R.

Robertson, B. E., Tacchella, S., Johnson, B. D., Hainline, K., Whitler, L., Eisenstein, D. J., Endsley, R., Rieke, M., Stark, D. P., Alberts, S., Dressler, A., Egami, E., Hausen, R., Rieke, G., Shivaei, I., Williams, C. C., Willmer, C. N. A., Arribas, S., Bonaventura, N., Bunker, A., Cameron, A. J., Carniani, S., Charlot, S., Chevallard, J., Curti, M., Curtis-Lake, E., D'Eugenio, F., Jakobsen, P., Looser, T. J., Lützgendorf, N., Maiolino, R., Maseda, M. V., Rawle, T., Rix, H. W., Smit, R., Übler, H., Willott, C., Witstok, J., Baum, S., Bhatawdekar, R., Boyett, K., Chen, Z., de Graaff, A., Florian, M., Helton, J. M., Hviding, R. E., Ji, Z., Kumari, N., Lyu, J., Nelson, E., Sandles, L., Saxena, A., Suess, K. A., Sun, F., Topping, M., & Wallace, I. E. B. 2023, Identification and properties of intense star-forming galaxies at redshifts z > 10. *Nature Astronomy*, 7, 611–621. ADS Bibcode: 2023NatAs...7..611R.

Rowland, L. E., Hodge, J., Bouwens, R., Piña, P. E. M., Hygate, A., Algera, H., Aravena, M., Bowler, R., da Cunha, E., Dayal, P., Ferrara, A., Herard-Demanche, T., Inami, H., van Leeuwen, I., de Looze, I., Oesch, P., Pallottini, A., Phillips, S., Rybak, M., Schouws, S., Smit, R., Sommovigo, L., Stefanon, M., & van der Werf, P. 2024, REBELS-25: Discovery of a dynamically cold disc galaxy at z = 7.31. *Monthly Notices of the Royal Astronomical Society*,. Publisher: OUP ADS Bibcode: 2024MNRAS.tmp.2235R.

Sanders, R. L., Shapley, A. E., Topping, M. W., Reddy, N. A., & Brammer, G. B. 2024, Direct T e-based Metallicities of z = 2-9 Galaxies with JWST/NIRSpec: Empirical Metallicity Calibrations Applicable from Reionization to Cosmic Noon. *The Astrophysical Journal*, 962, 24. Publisher: IOP ADS Bibcode: 2024ApJ...962...24S.

Sarkar, A., Chakraborty, P., Vogelsberger, M., McDonald, M., Torrey, P., Garcia, A. M., Khullar, G., Ferland, G. J., Forman, W., Wolk, S., Schneider, B., Bautz, M., Miller, E., Grant, C., & ZuHone, J. 2024,. Unveiling the Cosmic Chemistry: Revisiting the Mass-Metallicity Relation with JWST/NIRSpec at $4 < z < 10$. arXiv:2408.07974 [astro-ph].

Schaerer, D., Marques-Chaves, R., Barrufet, L., Oesch, P., Izotov, Y. I., Naidu, R., Guseva, N. G., & Brammer, G. 2022, First look with JWST spectroscopy: Resemblance among z □ 8 galaxies and local analogs. *A&A*, 665, L4. Publisher: EDP Sciences.

Schouws, S., Stefanon, M., Bouwens, R., Smit, R., Hodge, J., Labbé, I., Algera, H., Boogaard, L., Carniani, S., Fudamoto, Y., Holwerda, B. W., Illingworth, G. D., Maiolino, R., Maseda, M., Oesch, P., & van der Werf, P. 2022, Significant Dust-obscured Star Formation in Luminous Lyman-break Galaxies at z 7-8. *The Astrophysical Journal*, 928, 31. Publisher: IOP ADS Bibcode: 2022ApJ...928...31S.

Sommovigo, L., Ferrara, A., Pallottini, A., Dayal, P., Bouwens, R. J., Smit, R., da Cunha, E., De Looze, I., Bowler, R. A. A., Hodge, J., Inami, H., Oesch, P., Endsley, R., Gonzalez, V., Schouws, S., Stark, D., Stefanon, M., Aravena, M., Graziani, L., Riechers, D., Schneider, R., van der Werf, P., Algera, H., Barrufet, L., Fudamoto, Y., Hygate, A. P. S., Labbé, I., Li, Y., Nanayakkara, T., & Topping, M. 2022, The ALMA REBELS Survey: cosmic dust temperature evolution out to z 7. *Monthly Notices of the Royal Astronomical Society*, 513, 3122–3135. Publisher: OUP ADS Bibcode: 2022MNRAS.513.3122S.

Strandet, M. L., Weiss, A., De Breuck, C., Marrone, D. P., Vieira, J. D., Aravena, M., Ashby, M. L. N., Béthermin, M., Bothwell, M. S., Bradford, C. M., Carlstrom, J. E., Chapman, S. C., Cunningham, D. J. M., Chen, C.-C., Fassnacht, C. D., Gonzalez, A. H., Greve, T. R., Gullberg, B., Hayward, C. C., Hezaveh, Y., Litke, K., Ma, J., Malkan, M., Menten, K. M., Miller, T., Murphy, E. J., Narayanan, D., Phadke, K. A., Rotermund, K. M., Spilker, J. S., & Sreevani, J. 2017, ISM Properties of a Massive Dusty Star-forming Galaxy Discovered at z □ 7. *The Astrophysical Journal*, 842, L15. Publisher: IOP ADS Bibcode: 2017ApJ...842L..15S.

Tamura, Y., Mawatari, K., Hashimoto, T., Inoue, A. K., Zackrisson, E., Christensen, L., Binggeli, C., Matsuda, Y., Matsuo, H., Takeuchi, T. T., Asano, R. S., Sunaga, K., Shimizu, I., Okamoto, T., Yoshida, N., Lee, M. M., Shibuya, T., Taniguchi, Y., Umehata, H., Hatsukade, B., Kohno, K., & Ota, K. 2019, Detection of the Far-infrared [O III] and Dust Emission in a Galaxy at Redshift 8.312: Early Metal Enrichment in the Heart of the Reionization Era. *The Astrophysical Journal*, 874, 27. Publisher: IOP ADS Bibcode: 2019ApJ...874...27T.

Taylor, A. J., Barger, A. J., & Cowie, L. L. 2022, Metallicities of Five $z > 5$ Emission-line Galaxies in SMACS 0723 Revealed by JWST. *The Astrophysical Journal*, 939, L3. Publisher: IOP ADS Bibcode: 2022ApJ...939L...3T.

Torrey, P., Vogelsberger, M., Marinacci, F., Pakmor, R., Springel, V., Nelson, D., Naiman, J., Pillepich, A., Genel, S., Weinberger, R., & Hernquist, L. 2019, The evolution of the mass-metallicity relation in IllustrisTNG. *Monthly Notices of the Royal Astronomical Society*,. arXiv:1711.05261 [astro-ph].

Trager, S. C., Faber, S. M., Worthey, G., & González, J. J. 2000, The Stellar Population Histories of Early-Type Galaxies. II. Controlling Parameters of the Stellar Populations. *The Astronomical Journal*, 120, 165–188. Publisher: IOP ADS Bibcode: 2000AJ....120..165T.

Trump, J. R., Haro, P. A., Simons, R. C., Backhaus, B. E., Amorín, R. O., Dickinson, M., Fernández, V., Papovich, C., Nicholls, D. C., Kewley, L. J., Brunker, S. W., Salzer, J. J., Wilkins, S. M., Almaini, O., Bagley, M. B., Berg, D. A., Bhatawdekar, R., Bisigello, L., Buat, V., Burgarella, D., Calabrò, A., Casey, C. M., Ciesla, L., Cleri, N. J., Cole, J. W., Cooper, M. C., Cooray, A. R., Costantin, L., Ferguson, H. C., Finkelstein, S. L., Fujimoto, S., Gardner, J. P., Gawiser, E., Giavalisco, M., Grazian, A., Grogin, N. A., Hathi, N. P., Hirschmann, M., Holwerda, B. W., Huertas-Company, M., Hutchison, T. A., Jogee, S., Juneau, S., Jung, I., Kartaltepe, J. S., Kirkpatrick, A., Koekemoer, A. M., Lotz, J. M., Lucas, R. A., Magnelli, B., Matharu, J., Pérez-González, P. G., Pirzkal, N., Rafelski, M., Rose, C., Seillé, L.-M., Somerville, R. S., Straughn, A. N., Tacchella, S., Vanderhoof, B. N., Weiner, B. J., Wuyts, S., Yung, L. Y. A., & Zavala, J. A. 2022,. The Physical Conditions of

Emission-Line Galaxies at Cosmic Dawn from JWST/NIRSpec Spectroscopy in the SMACS 0723 Early Release Observations. arXiv:2207.12388.

Ucci, G., Dayal, P., Hutter, A., Kobayashi, C., Gottlöber, S., Yepes, G., Hunt, L., Legrand, L., & Tortora, C. 2023, Astraeus V: the emergence and evolution of metallicity scaling relations during the epoch of reionization. *Monthly Notices of the Royal Astronomical Society*, 518, 3557–3575. Publisher: OUP ADS Bibcode: 2023MNRAS.518.3557U.

Wilkins, S. M., Vijayan, A. P., Lovell, C. C., Roper, W. J., Zackrisson, E., Irodotou, D., Seeyave, L. T. C., Kuusisto, J. K., Thomas, P. A., Caruana, J., & Conselice, C. J. 2023, First Light And Reionization Epoch Simulations (FLARES) VII: The star formation and metal enrichment histories of galaxies in the early Universe. *Monthly Notices of the Royal Astronomical Society*, 518, 3935–3948. Publisher: OUP ADS Bibcode: 2023MNRAS.518.3935W.

The First Chapters of Our Cosmic History with JWST
Proceedings IAU Symposium No. 391, 2025
T. Nanayakkara & M. Maseda, eds.
doi:10.1017/S1743921324002990

Probing galaxy build-up at z>3: NIRCAM and MIRI synergy in EGS

Belén Alcalde Pampliega

European Southern Observatory, Chile
email: Belen.Alcalde@eso.org

Abstract. JWST is at the forefront of revolutionizing our comprehension of the high-redshift Universe, stretching the boundaries of galaxy detection. Yet, as we strive for a more comprehensive understanding of the high-redshift Universe, concerns arise about the consistency of early estimates of galaxy number density, their stellar masses, and physical properties.

Our study delves into the synergy of NIRCAM and MIRI in the Extended Groth Strip (EGS) field, unveiling insights into the characterization of approximately 300 galaxies at z>3, including 30 MIRI-detected red massive galaxies selected as HST-dark/faint and Little Red Dots (LRDs). By providing constraints on rest-frame optical and near-infrared emission, MIRI allows for more accurate measurements of stellar mass and helps disentangle the contribution from star formation and AGN activity. Although MIRI data has minimal impact on overall photometric redshift accuracy, it is critical for reliable stellar mass determination, especially at the highest redshifts (z>7). Excluding MIRI data can lead to overestimating stellar mass by up to 0.15 dex for the general population and 0.4-1 dex for LRDs. Besides, MIRI data significantly improve our ability to constrain the contribution of Active Galactic Nuclei (AGN) to the observed light. For red massive galaxies, or those with the longest NIRCam filters contaminated by strong emission lines, excluding MIRI data can lead to a substantial overestimation of the AGN fraction.

Keywords. Galaxies: high-redshift, stellar content, active, fundamental parameters; Infrared: galaxies

1. Introduction

A pressing challenge concerning the early assembly of massive galaxies did already exist in the pre-JWST era, as revealed by their number density and maximum stellar mass, which seems to affect massive red galaxies more drastically (AlcaldePampliega et al. 2019; Wang et al. 2019). The advent of JWST, has not only confirmed their nature (e.g., PerezGonzalez et al. 2023; Barrufet et al. 2023) and constrained their number densities (e.g., Gottumukkala et al. 2024), but it has also allowed for the discovery of very massive galaxies up to redshift 9 (e.g., $> 10^{10}\mathrm{M}_{\odot}$, Labbe et al. 2023a) and above (e.g., Harikane et al. 2024). Additionally, JWST has unveiled a new galaxy population, the so-called Little Red Dots (LRDs), characterized by v-shaped spectra with blue rest-UV, and red rest-optical colors which could be dominated by the light arising from the dusty torus of the AGN (e.g., Killi et al. 2023; Greene et al. 2024).

The build-up of such massive systems, in a few hundred Myrs after the Big Bang, poses a significant challenge to ΛCDM cosmology. A key limitation in pre-JWST and NIRCAM-only studies has been the substantial uncertainty in derived stellar masses due to the lack of rest-frame near-infrared data, which is crucial for precise mass estimates. Beyond cosmic noon (i.e., $z\gtrsim3$), MIRI covers the rest-frame optical and near-infrared and key strong emission lines, essential to constrain the masses of red galaxies (e.g., Williams et al. 2024; Wang et al. 2024) and even to quantify the contribution from AGNs

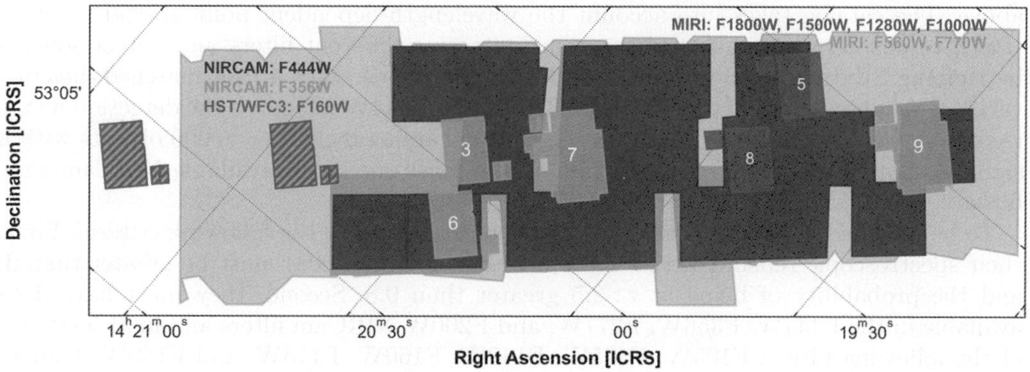

Figure 1. Layout of the CANDELS/HST and CEERS/JWST observations in the EGS field. The region covered by the HST/WFC3 F160W filter is shown in grey, while the JWST/NIRCAM F356W and F444W are shown in orange and red. MIRI pointings (white numbers) from short (F560W and F770W) and long (F1500W, F1280W, F1500W, and F1800W) wavelength filters are colored in green and blue respectively. We note that only 2 MIRI pointings are covered by all MIRI filters, but none of them overlap with NIRCAM footprints.

(e.g., Barro et al. 2024). In this work, we take advantage of the MIRI data available in the Extended Groth Strip (EGS) field to quantify its relevance to further constrain the SEDs of z>3 galaxies, and improve the reliability of the derived physical properties with a targeted focus on their redshift, stellar mass, and AGN contribution.

2. Data

Here, we focus on the EGS field covered by both HST and the Cosmic Evolution Early Release Science (CEERS) JWST survey. Specifically, we use data from the early data releases of the NIRCam and MIRI instruments together with HST archival data. The HST and NIRCam data used in this study were made publicly available by G. Brammer† and reduced using the `Grizli` pipeline‡. The images include all available HST and JWST observations over the CEERS field of view. In particular, NIRCam data (F115W, F150W, F200W, F277W, F356W, F410M, and F444W filters) ranges from 1.15 μm to 4.44 μm. HST data includes ancillary data from ACS (F435W, F606W, and F814W) and WFC3 (F105W, F125W, F140W and F160W).

We also use the MIRI images made publicly available by the CEERS data release 0.6 (Yang et al. 2023b)§. The CEERS survey has eight MIRI pointings, however, in this work, we focus only on the 6, the 'short' (3, 6, 7 and 9) and 'long' (5 and 8), overlapping with the NIRCam footprint (Fig. 1). F560w and/or F770W filters are used for the 'short' pointings, while the 'long' ones are covered by at least one of the following: F1000W, 1280W, 1500W, and 1800W. Thus, although CEERS NIRCam pointings comprise $\sim$ 100 arcmin^2, also covered by the HST/CANDELS survey (Grogin et al. 2011; Koekemoer et al. 2011), we focus on the areas covered by both NIRCam and MIRI simultaneously. This represents a total area of $\sim$15 arcmin^2 ($\sim$10 and 5 arcmin^2 for the short and long MIRI filters respectively).

3. Sample selection

The galaxies presented in this work belong to a PSF-matched photometric catalog built using both HST and JWST/NIRCam available imaging data in the CEERS field listed

† `https://github.com/gbrammer/grizli/blob/master/docs/grizli/image-release-v6.rst`
‡ `https://github.com/gbrammer/grizli`
§ `https://ceers.github.io/dr06.html#miri-imaging`

above. The catalog takes into account the wavelength-dependent point-spread function (PSF) to ensure accurate flux measurements across different filters and detect sources by running SExtractor (Bertin et al. 1996) in dual mode, with a PSF-matched, inverse-variance weighted stack of the F277W, F356W, and F444W images as the detection image (see Weibel et al. 2024, for details ¶). The initial catalog includes ~85000 objects with a signal-to-noise ratio (SNR) greater than 5 in at least one of the available NIRCam-wide filters.

To be included in the z>3 catalog, objects must also meet the following criteria. First, their spectroscopic redshift (`from EAZY,` Brammer et al. 2008) must be greater than 3, and the probability of being at z>2.5 greater than 0.8. Second, they must have data available in the F444W, F356W, F277W, and F200W NIRCam filters and in at least two of the following filters: F105W, F125W, F140W, F160W, F115W, and F150W. Fourth, the SNR in the detection image (F277W, F356W, and F444W stack) must be greater than 8.5. Finally, the objects must not be flagged as stars or artifacts. The derived z>3 catalog contains ~10400 galaxies.

3.1. *MIRI detected z>3 galaxies*

Using Weibel et al. (2024) catalog as a parent sample, we select all sources that overlap with any CEERS MIRI pointings, leading to a sample of 1931 galaxies and force photometry in all available MIRI images. To ensure consistent photometry along the whole wavelength range, we adopt the same photometric approach followed for the parent catalog. Briefly, we measure MIRI fluxes in circular apertures with a radius of $0.16''$ at the position of the source in the F277W+F356W+F444W detection image, and we apply the same aperture correction as in the NIRCam/F444W PSF-matched parent catalog. Additionally, given MIRI's much broader and different PSFs, we apply a PSF correction based on the encircled energy ratio (computed using `webbpsf`) between each MIRI filter and NIRCam/F444W filter. We then select objects with an SNR greater than 1(3) in at least one MIRI filter. This leads to a total of 644(148) galaxies, 463(136), and 181(12) covered by 'short' and 'long' MIRI filters respectively.

Several color selection criteria have been proposed (Wang et al. 2016; AlcaldePampliega et al. 2019; Barrufet et al. 2023; PerezGonzalez et al. 2023) to identify a population of massive red galaxies missed by most pre-JWST selections, due to their faintness in the optical-to-NIR regime, the so-called HST-dark galaxies. Fig. 2, shows our sample within two of those color selection diagrams: F150W-F356W vs. F356W (*Left*) and F150W-F444W vs. F444W (*middle*). The galaxies that fulfill the color criteria (F150W-F356W> 1.5 mag or F150W-F444W> 2 mag) and a magnitude cut (F150W> 23 mag) but are not included in the z>3 catalog are shown by orange and red circles, respectively. Additionally, the right panel from Fig. 2 shows one of the color criteria used to select LRDs (F150-F200W< 0.5 mag & F227-F444W> 1.5 mag, PerezGonzalez et al. 2024). Only 3 of the MIRI-detected galaxies qualify as LRDs according to this diagram.

Not surprisingly, while only 8% of the MIRI-covered galaxies have a SNR> 3 detection in at least one MIRI filter, percentages are much higher, 80%(F150W-F356W> 1.5 mag) and 96%(F150W-F444W> 2 mag), for red galaxies. Additionally, three LRDs are detected in MIRI (SNR> 3).

4. SED fitting and redshift accuracy

Obtaining reliable physical properties of galaxies requires knowing the distance to those sources, either by spectroscopic measurements or by high-quality photometric redshifts.

¶ `https://github.com/weibelan/W24-SMF-paper-catalogs`

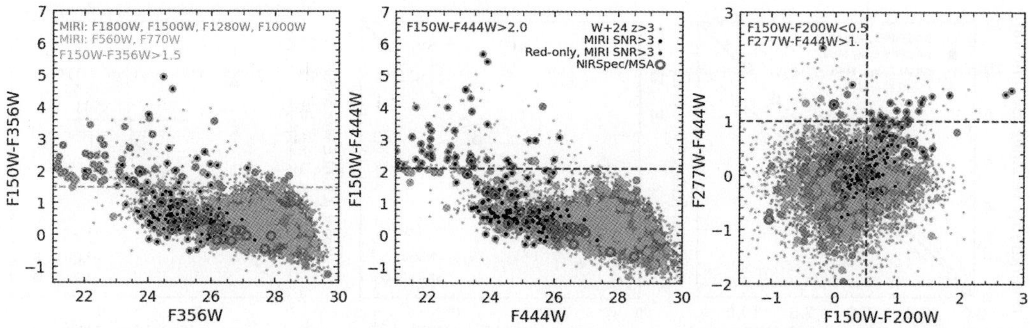

Figure 2. Our sample within color-magnitude diagrams showing color criteria proposed to select massive red galaxies at high-z (z$\gtrsim$3). From left to right the F150W-F356W>1.5 mag, F150W-F444W>2 mag, and F150W-F200W<0.5 mag and F277W-F444W>1 mag. Grey dots show the HST+NIRCAM z>3 CEERS sample, black dots show those covered by any MIRI filter and detected at SNR>3. The orange and red dots represent the MIRI-detected galaxies not included in the z>3 catalog and fulfilling the F150W-F356W>1.5, and F150W-F444W>2 color selections. Galaxies covered by MIRI short (F560W and F770W) or long wavelengths (F1000W, F1280W, F1500W, and F1800W) are highlighted in green and blue respectively. Spectroscopically confirmed sources are encircled in grey. We note that, due to the CEERS survey setup, lower and higher wavelength MIRI filters cover different footprints in the field (see Fig. 1).

Most recent works on large samples of high-z galaxies found by JWST are mainly based on NIRCam broad-band photometry, combined with HST photometry when available, due to the limited availability of both medium-band photometry and MIRI data. Here, we test the relevance of MIRI data in constraining the photometric redshifts of z>3 galaxies. As a first step, we quantify the quality of photometric redshift using `CIGALE` (Boquien et al. 2019) and `BAGPIPES` (Carnall et al. 2018) for the subsample with available spectroscopic redshifts. Then, we compare the results for the whole sample. We explore a similar parameter space for both codes in the redshift range $z = (0, 15)$. Briefly, we assume a delayed-τ SF history with $\tau = (0.1, 9)$ Gyr, masses in the range $\log \mathrm{M}_\star/\mathrm{M}_\odot = (6, 13)$, a Calzetti dust model with $\mathrm{A_V} = (0.0, 4.0)$ mag, nebular emission with an ionization parameter of $\log \mathrm{U} = -2$, a velocity dispersion of 300 km/s and a metallicity between $\mathrm{Z} = (0.2, 1.)\ \mathrm{Z}_\odot$.

Overall, we find a great agreement (see Fig. 3) and very few outliers in the derived z with and without MIRI, both for spectroscopically confirmed galaxies ($\sigma_{NMAD} \sim 0.02 - 0.08$) and our photometric sample (middle and right panels in Fig. 3), regardless of the code used. This implies that adding a few MIRI filters can only slightly improve the photometric redshift, as also found by (Li et al. 2024) for z<3 galaxies. Thus, for the rest of the work, we fix the redshift to that derived using also MIRI, and we focus on the results obtained from `CIGALE` simultaneously computing the AGN contribution (fAGN; from 0 to 0.99 over the rest-frame 3-30μm wavelength range).

5. Stellar mass estimates

The left panel in Fig. 4 shows the difference in the derived $\mathrm{M}_\star$ as a function of redshift for the subsamples presented in the previous section. For the general population, dominated by $\mathrm{z} \sim 3 - 5$ intermediate mass ($\log\mathrm{M}_\star \sim 8 - 10\mathrm{M}_\odot$) galaxies, we do not find a significant difference in the derived stellar mass. However, for the highest redshifts (z$\gtrsim$7), the derived stellar masses are overestimated by $\sim$0.15 dex when excluding MIRI/F560 and/or F770W filters.

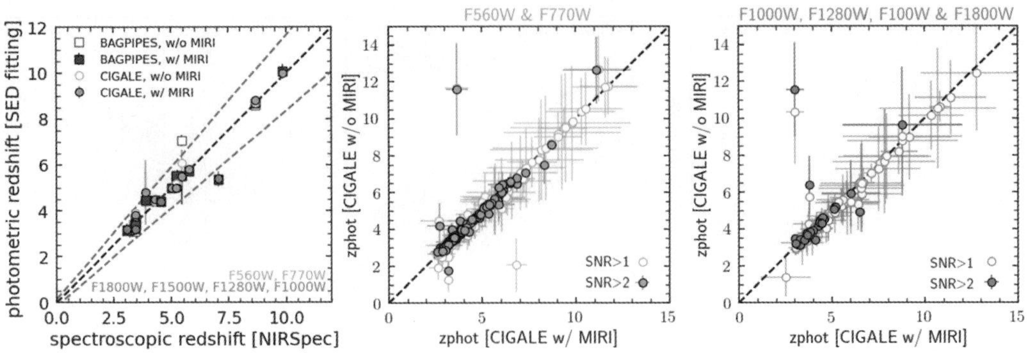

Figure 3. *Left:* Photometric versus spectroscopic redshift obtained both with (filled symbols) and without (empty symbols) MIRI using BAGPIPES (squares) and CIGALE (circles). Galaxies with coverage in the shorter or longer filters are shown in green and blue, respectively. *Middle and right:* Comparison of the photometric redshift obtained for the wholes sample with and without MIRI for the short (*middle*) and long (*right*) MIRI filters. Low and high SNR data are shown with empty and filled symbols, respectively.

Our sample of red galaxies behaves similarly to the general population, with only a slight decrease (∼0.1-0.2 dex) of the stellar mass for the highest redshifts and the shorter MIRI/F560W and F770 filters. Wang et al. (2024) also reported a decrease in $M_\star$ with redshift (<0.2 dex at z<6 and ∼0.4 dex at z>6.5) for a larger sample of red galaxies in PRIMER. Similarly, Williams et al. (2024) reported a median decrease of 0.6 dex for red z>3 sources with high masses ($\log M_\star > 10$). Finally, when focusing on the LRDs, we find a much larger discrepancy ($\sim 0.4-1$ dex). While our statistics are not significant for this subsample, Akins et al. (2024) reports a similar trend (i.e., systematic offsets of $\sim 1-2$ dex, towards larger masses when not including the MIRI/F770W filter) for a large sample of LRDs. We note that the most common cause for stellar mass overestimation (using HST+NIRCam broad-band photometry) is the contamination of the longest available NIRCam filter by the presence of strong emission lines. Specifically, Hβ+[OIII] (one of the brightest lines in normal star-forming galaxies) lies within F444W at z$\simeq 7-8.8$; the redshift range where the difference becomes more relevant. Briefly, our results demonstrate the significance of MIRI data for accurately determining the stellar mass of high redshift galaxies. This is particularly relevant as it affects the stellar mass that distant galaxies are able to assemble at early epochs and mitigates the tension with expectations for the halo mass function in ΛCDM cosmology.

6. Constraining the AGN contribution

At z>3, while most PAH features are not any more accessible, MIRI can still help quantify the AGN faction by constraining the fluxes at longer wavelengths. Right Panel in Fig. 4 shows the difference in the derived AGN fraction (fAGN) as a function of $M_\star$ and highlights how the absence of MIRI data can lead to critical overestimates (up to 80%) of the AGN contribution. We note that, while for the majority of the galaxies, the fAGN does not vary and stays within a negligible value ($\lesssim$0.1), for most of the galaxies with a derived AGN contribution using HST and NIRCam data only, the fAGN substantially decreases when adding MIRI data. This effect is more relevant for massive galaxies, including red galaxies and LRDs. This can be caused by their characteristic red colors as seen by the longest NIRCAM filters. The lack of constraints at longer wavelengths allows the templates to continue rising while, as proved by our results, those templates are often inconsistent with the measured MIRI photometry. After examining the best SED fit, both with and without MIRI, we find that this trend is caused by

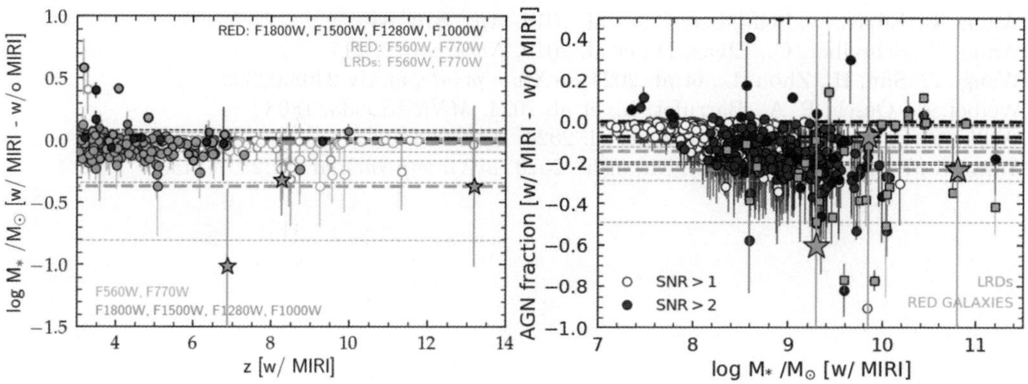

Figure 4. *Left:* Differences in the $M_\star$ estimation with and without MIRI data as a function of redshift. Galaxies covered by MIRI short or long wavelengths are shown in green and blue for the general population and orange and red for red galaxies. LRDs are highlighted with pink stars. Empty symbols correspond to galaxies with SNR<2 in MIRI filters. The 50th percentile is shown with a dashed thick line and the 16th and 84th percentiles are shown with thin lines. *Right:* Differences in the derived AGN fraction with and without MIRI data as a function of $M_\star$. Short or long wavelengths are shown in red and purple respectively.

the difference between the SED-predicted MIRI photometry when excluding MIRI for fit, and the real photometry. Indeed, the median difference increases for larger distances between the last available NIRCam filter (often F444W) and the first available MIRI filter (from $< 0.5\mu$Jy in F560W to $\sim 6\mu$Jy in F1800W). Outliers, however, can go from $< 4\mu$Jy in F560W to $> 15\mu$Jy in F1800W. Besides, we also emphasize the crucial role of upper limits in constraining the AGN contribution in all but most heavily red galaxies; either intrinsic or due to the presence of an emission line boosting the flux of the last available NIRCam filter.

These results, demonstrate the effectiveness of MIRI to provide strong constraints on mid-IR luminosities from both star formation and AGN, facilitating studies on galaxy/AGN co-evolution.

References

Akins, H. B., Casey, C. M., Lambrides, E., et al. 2024, *arXiv e-prints*, arXiv:2406.10341
Alcalde Pampliega, B., Pérez-González, P. G., Barro, G., et al. 2019, *ApJ*, 876, 135
Barro, G., Pérez-González, P. G., Kocevski, D. D., et al. 2024, *ApJ*, 963, 128
Barrufet, L., Oesch, P. A., Weibel, A., et al. 2023, *MNRAS*, 522, 449
Bertin, E. & Arnouts, S. 1996, *Aaps*, 117, 393
Boquien, M., Burgarella, D., Roehlly, Y., et al. 2019, *Aap*, 622, A103
Brammer, G. B., van Dokkum, P. G., & Coppi, P. 2008, *ApJ*, 686, 1503
Carnall, A. C., McLure, R. J., Dunlop, J. S., & Davé, R. 2018, *MNRAS*, 480, 4379
Gottumukkala, R., Barrufet, L., Oesch, P. A., et al. 2024, *MNRAS*, 530, 966
Greene, J. E., Labbe, I., Goulding, A. D., et al. 2024, *ApJ*, 964, 39
Grogin, N. A., Kocevski, D. D., Faber, S. M., et al. 2011, *ApJS*, 197, 35
Harikane, Y., Inoue, A. K., Ellis, R. S., et al. 2024, *arXiv e-prints*, arXiv:2406.18352
Killi, M., Watson, D., Brammer, G., et al. 2023, *arXiv e-prints*, arXiv:2312.03065
Koekemoer, A. M., Faber, S. M., Ferguson, H. C., et al. 2011, *ApJS*, 197, 36
Labbe, I., Greene, J. E., Bezanson, R., et al. 2023, *arXiv e-prints*, arXiv:2306.07320
Labbé, I., van Dokkum, P., Nelson, E., et al. 2023, *Nature*, 616, 266
Li, Q., Conselice, C. J., Adams, N., et al. 2024, *MNRAS*, 531, 617
Pérez-González, P. G., Barro, G., Annunziatella, M., et al. 2023, *ApJL*, 946, L16
Pérez-González, P. G., Barro, G., Rieke, G. H., et al. 2024, *ApJ*, 968, 4

Wang, T., Elbaz, D., Schreiber, C., et al. 2016, *ApJ*, 816, 84
Wang, T., Schreiber, C., Elbaz, D., et al. 2019, *Nature*, 572, 211
Wang, T., Sun, H., Zhou, L., et al. 2024, *arXiv e-prints*, arXiv:2403.02399
Weibel, A., Oesch, P. A., Barrufet, L., et al. 2024, *MNRAS*, 533, 1808
Williams, C. C., Alberts, S., Ji, Z., et al. 2024, *ApJ*, 968, 34
Yang, G., Papovich, C., Bagley, M., et al. 2023, *arXiv e-prints*, arXiv:2307.14509

The First Chapters of Our Cosmic History with JWST
Proceedings IAU Symposium No. 391, 2025
T. Nanayakkara & M. Maseda, eds.
doi:10.1017/S1743921324002308

Divide et Impera: Galaxy Assembly in the Early Universe from NIRSpec/IFU Observations of ALMA/REBELS Galaxies

Mauro Stefanon
Departament d'Astronomia i Astrofisica, Universitat de València, C. Dr. Moliner 50, E-46100 Burjassot, València, Spain
Unidad Asociada CSIC "Grupo de Astrofísica Extragaláctica y Cosmología" (Instituto de Física de Cantabria - Universitat de València)
email: mauro.stefanon@uv.es

Abstract. The discovery in JWST data of numerous bright galaxies at Cosmic Dawn, along with the identification of potentially "Universe-breaking" massive galaxies, is reshaping our understanding of galaxy formation, and raise questions about how rapidly galaxies assembled their mass in the early Universe. In parallel with follow-up studies, significant progress can be achieved by characterizing the brightest and most massive galaxies observed when the universe was 200 − 400 million years older, likely the most immediate descendants of the tantalizing objects observed at z>10. In this talk, we will present the first results from a NIRSpec IFU R100 program targeting 12 bright, especially high-mass galaxies at just approximately 750Myr of cosmic time (6.5<z<8.5). Remarkably, this sample also benefits from [CII] and dust continuum estimates from the recent ALMA large program REBELS, making this a unique sample in the panorama of high-redshift studies. Leveraging the JWST's high spatial and spectral resolution across a broad wavelength range, from the rest-frame UV to rest-frame optical, we can reconstruct spatially-resolved maps of their stellar and gas content. These maps enable us to interpret them in terms of in-situ mass assembly, merger-driven star formation, and star-formation history. We will also provide a brief outlook on their formation histories as inferred from gas-phase metallicities, but will leave most of the discussion to a companion, more in-depth presentation. These results will ultimately allow us to assess the star formation efficiency in massive galaxies at early epochs.

Keywords. galaxies: high-redshift; galaxies: ISM; galaxies: evolution; galaxies: star formation

1. Introduction

In the pre-JWST era, a number of studies indicated that galaxies assembled their stellar content at the same rate at which the host halos accreted dark matter (e.g., Oesch et al. 2018; Stefanon et al. 2021). The unprecedented JWST/NIRCam resolution and sensitivity at near-infrared wavelengths has opened a new window into our knowledge of the early Universe. One of the most surprising result has been the discovery of numerous candidate luminous galaxies at $z \gtrsim 10$ (e.g., Naidu et al. 2022; Harikane et al. 2023; Bradley et al. 2023; Castellano et al. 2023; Finkelstein et al. 2023; Atek et al. 2023; Adams et al. 2023; Donnan et al. 2023; Rodighiero et al. 2023; Casey et al. 2024). Moreover, the much improved spatial resolution and sensitivity provide us with a solid probe into the rest-frame optical emission of galaxies at early cosmic times, allowing us to better constrain their stellar populations. Remarkably, already the first batch of NIRCam observations showed the existence of exceptionally massive ($M_\star > 10^{10} M_\odot$) galaxies at $z \gtrsim 8$ (e.g., Labbé et al. 2023). These results suggest a very vital Universe during the first stages of

Figure 1. Illustration of the variety of physical processes in massive star-forming galaxies at $z \sim 7$. Represented are false-color stamps of five sources from the REBELS sample with IFU coverage, constructed combining the Hα, rest-frame optical and rest-frame UV light. The diameter of the stamp is about 6 kpc.

galaxy assembly. However, even though the existence of these bright sources is predicted by models (e.g., Mason et al. 2015; Mashian et al. 2016; Trac et al. 2015; Williams et al. 2018), their prevalence is expected to be much lower than observed.

2. A unique sample of exceptionally bright, star-forming galaxies at $z \sim 7$ with ALMA and JWST/IFU observations

Significant progress in understanding early galaxy formation can be achieved by studying sources at slightly later epochs, complementing current efforts to probe the highest redshifts. The most promising targets for exploring galaxy assembly in the early Universe are the brightest and most massive galaxies identified so far. The Re-ionization Era Bright Emission Line Survey (REBELS – Bouwens et al. 2021) is a cycle 7 ALMA large program targeting 40 exceptionally bright ($M_{\rm UV} \sim -22$ mag) galaxies candidate to be at $6.5 \lesssim z \lesssim 8.5$. These galaxies were identified across 6.5 square degrees of data with the deepest optical, NIR and Spitzer observations, mainly in the COSMOS/UltraVISTA and VIDEO/XMM fields. Through spectral scans for [CII]$\lambda 158\mu$m, REBELS has provided spectroscopic confirmation of 26 massive galaxies at $6.5 < z < 7.5$, including dust emission detection in 18 of them. The large masses and luminosities of these samples are consistent with being descendant of the exceptionally bright galaxies recently identified at $z \gtrsim 10$, making them unique laboratories to study the efficiency of galaxy assembly in the early Universe. Comparison between the spatial distribution of [CII], dust continuum, and rest-frame UV light reveals a great diversity of physical conditions: massive dusty neighbours which had remained hidden from the rest-UV selection (e.g., Fudamoto et al. 2021), offsets between the emission of dust continuum and that from [CII] and rest-UV, clumpy morphologies, merger events, gradients in the spatial distribution of dust continuum and [CII] emission, indications of potential outflows (not shown).

Notably, 12 of these galaxies were recently observed in JWST Cycle 1 GO program 1626 (PI: Stefanon), with 30-minute NIRSpec/IFU Prism integrations capturing spectra across the $0.9 - 5.3\mu$m range (rest-frame UV to Hα). Figure 1 displays false-color images for 5 of these sources, created by combining Hα, rest-frame optical, and rest-frame UV light, revealing a diverse array of morphologies and evolutionary stages within each galaxy's footprint.

Figure 2 illustrates the depth of information accessible from the IFU cubes. Here we briefly introduce some of the analyses enabled by this dataset. Thanks to the simultaneous coverage of Hα and Hβ we can construct spatially-resolved maps of dust extinction from the Balmer decrement, while the detection of [OII]λ3727 and of the [OIII] doublet at $\lambda \sim 5000$Å enable gas-phase metallicity measurements. By comparing rest-frame UV emission, which traces star formation in the last ~ 200Myr (e.g., Kennicutt and Evans

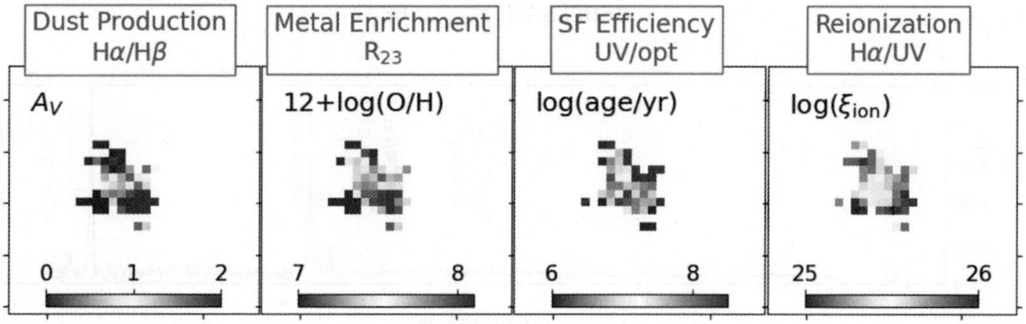

Figure 2. Illustration of the diversity of information provided by the IFU data. Each panel refers to a different physical observable, as labelled at the top. These maps indicate a complex interaction of physical mechanisms that are simultaneously occurring in massive $z \sim 7$ galaxies at the epoch of observation and in their past, without any single process being predominant.

2012), with rest-frame optical light that primarily reflects the more evolved stellar populations (a proxy for stellar mass), we can estimate how efficiently these galaxies assembled their stellar content. Additionally, we can probe the intensity of the ionizing field by comparing the intensity of the Hα emission to that of the rest-frame UV (e.g. Bouwens et al. 2016). Most notably, these maps suggest a complex interplay of physical mechanisms concurrently ongoing across these galaxies at the epoch of the observations and in their past, with no dominant process.

A compelling example of the JWST-ALMA synergy's potential is one of our sample sources, where the rest-frame UV emission is spatially complementary to the dust emission, indicating that approximately 90% of the UV light is obscured by dust. The picture becomes even more intriguing when examining the morphology of rest-frame optical emission, revealing a significant ($\sim 50\%$) stellar population hidden behind the dust. This finding aligns with recent studies identifying massive ($M_\star > 10^{11} M_\odot$) dust-obscured galaxies at $z > 4$ (e.g., Xiao et al. 2023), which contribute substantially to the cosmic stellar mass at these redshifts.

3. A massive $Z_\odot$ galaxy in the Epoch of Reionization

Recent studies have unveiled the existence of massive galaxies with evolved stellar populations up to $z \sim 8$ (e.g., Labbé et al. 2023; Carnall et al. 2023; Glazebrook et al. 2024; Nanayakkara et al. 2024; de Graaff et al. 2024; Weibel et al. 2024). Additionally, the detection of [OIII]88μm with ALMA (Schouws et al. 2024; Carniani et al. 2024a) in GS-z14 – the current highest-redshift spectroscopically confirmed galaxy (Carniani et al. (2024b)) – hints at ISM enrichment occurring at early cosmic times. However, current metallicity estimates for statistically significant samples of galaxies at $4 \gtrsim z \gtrsim 10$ suggest low metallicities ($\sim 10-20\% Z_\odot$ – e.g., Curti et al. 2023; Nakajima et al. 2023). These results are based on faint/low mass samples, leaving open questions about the chemical enrichment of the ISM in brighter/more massive galaxies.

Fortunately, thanks to the IFU observations for the 12 REBELS galaxies, we are now in a position to explore the metallicity of more massive galaxies at $z \sim 7$. Our measurements indicate on average higher metallicities than estimates in the literature at similar redshifts, although with a large ($\sim$ 1dex) dispersion. These results were presented in a dedicated talk in this Symposium (#2711 – L. Rowland: *JWST NIRSpec IFU observations of metal-rich, massive galaxies in the Epoch of Reionisation*) which we refer to for more details.

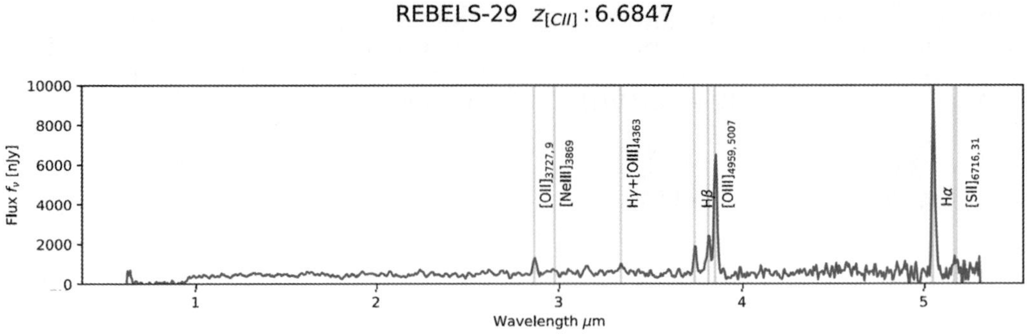

Figure 3. Spectrum, in f_nu units, of the REBELS source with solar metallicity. The main emission lines are indicated by the vertical dotted lines and labelled accordingly. Our analysis, based on the resolved stellar population properties, indicates that this galaxies started to form its stars at $z > 9$.

Among the REBELS sample, one galaxy stands out for its $Z \sim Z_\odot$ metallicity. Figure 3 presents its spectrum, revealing intense [OIII] and [OII] line emission, and potential [SII] emission, further supporting the high metallicity value we found. To gain insight on how rapidly this galaxy could have formed, we leveraged the IFU cube's spatial data to map the stellar population age. Specifically, we calculated the ratio between the Hα line emission and the corresponding rest-frame optical continuum obtained by combining the flux density between 4000Å and 6500Å rest-frame, after masking the emission lines. This ratio approximates the specific star-formation rate (sSFR), which inversely correlates with stellar population age. The resulting map presents a clear gradient across the galaxy, indicative of different timescales of stellar mass assembly. An analysis using standard SED fitting techniques of the spectrum resulting from the combination of the spaxels with the lowest sSFR (i.e., possibly corresponding to the older populations) results in a stellar population age between 200 and 750 Myr, suggesting that this galaxy started to form its stars at an epoch corresponding to $z > 9$.

4. Conclusions

JWST has opened the window to the exploration of even higher redshifts than before, well into the Cosmic Dawn Era. Most importantly, it is enabling transitioning from global (average) population properties to the physical characterization of individual sources. The complementary nature of JWST's spatially-resolved optical data and ALMA's dust continuum measurements enables a comprehensive analysis of the galaxy's assembly history, demonstrating the power of combining these observatories for studying high-redshift galaxies. Thanks to NIRSpec IFU/Prism observations for a sample of 12 bright star forming galaxies at $z \sim 7$ with ALMA [CII] and dust continuum measurements from the REBELS program, we have been able to characterize their ISM chemical enrichment. In particular we have identified one galaxy with a solar metallicity. Leveraging the high spatial and spectral resolution enabled by the IFU observations, we estimate that this galaxy started to form its stars at an epoch corresponding to $z > 9$. This is consistent with scenarios where massive galaxies in the early universe experience intense starburst phases, which drive rapid growth in stellar mass and metallicity.

A forthcoming JWST C3 program (PI: Schouws) will obtain NIRCam Imaging and WFSS observations of 25 [CII]-luminous, massive star-forming galaxies at $z \sim 7$ (including the 12 galaxies considered for this study), enabling the characterization of their environment, their dark matter halos and the identification of ionized bubbles.

References

Adams, N. J., Conselice, C. J., Ferreira, L., Austin, D., Trussler, J. A. A., Juodžbalis, I., Wilkins, S. M., Caruana, J., Dayal, P., Verma, A., & Vijayan, A. P. 2023, Discovery and properties of ultra-high redshift galaxies (9 < z < 12) in the JWST ERO SMACS 0723 Field. *MNRAS*, 518(3), 4755–4766

Atek, H., Chemerynska, I., Wang, B., Furtak, L. J., Weibel, A., Oesch, P., Weaver, J. R., Labbé, I., Bezanson, R., van Dokkum, P., Zitrin, A., Dayal, P., Williams, C. C., Nannayakkara, T., Price, S. H., Brammer, G., Goulding, A. D., Leja, J., Marchesini, D., Nelson, E. J., Pan, R., & Whitaker, K. E. 2023, JWST UNCOVER: discovery of z > 9 galaxy candidates behind the lensing cluster Abell 2744. *MNRAS*, 524(4), 5486–5496

Bouwens, R. J., Smit, R., Labbé, I., Franx, M., Caruana, J., Oesch, P., Stefanon, M., & Rasappu, N. 2016, The Lyman-Continuum Photon Production Efficiency ξ_{ion} of z ∼ 4-5 Galaxies from IRAC-based Hα Measurements: Implications for the Escape Fraction and Cosmic Reionization. *ApJ*, 831(2), 176

Bouwens, R. J., Smit, R., Schouws, S., Stefanon, M., Bowler, R., Endsley, R., Gonzalez, V., Inami, H., Stark, D., Oesch, P., Hodge, J., Aravena, M., da Cunha, E., Dayal, P., de Looze, I., Ferrara, A., Fudamoto, Y., Graziani, L., Li, C., Nanayakkara, T., Pallotini, A., Schneider, R., Sommovigo, L., Topping, M., van der Werf, P., Algera, H., Barrufet, L., Hygate, A., Labbe, I., Riechers, D., & Witstok, J. 2021, Reionization Era Bright Emission Line Survey: Selection and Characterization of Luminous Interstellar Medium Reservoirs in the z>6.5 Universe. *arXiv e-prints*, arXiv:2106.13719

Bradley, L. D., Coe, D., Brammer, G., Furtak, L. J., Larson, R. L., Kokorev, V., Andrade-Santos, F., Bhatawdekar, R., Bradač, M., Broadhurst, T., Carnall, A., Conselice, C. J., Diego, J. M., Frye, B., Fujimoto, S., Hsiao, T. Y. Y., Hutchison, T. A., Jung, I., Mahler, G., McCandliss, S., Oguri, M., Postman, M., Sharon, K., Trenti, M., Vanzella, E., Welch, B., Windhorst, R. A., & Zitrin, A. 2023, High-redshift Galaxy Candidates at $z = 9 - 10$ as Revealed by JWST Observations of WHL0137-08. *ApJ*, 955(1), 13

Carnall, A. C., McLure, R. J., Dunlop, J. S., McLeod, D. J., Wild, V., Cullen, F., Magee, D., Begley, R., Cimatti, A., Donnan, C. T., Hamadouche, M. L., Jewell, S. M., & Walker, S. 2023, A massive quiescent galaxy at redshift 4.658. *Nature*, 619(7971), 716–719

Carniani, S., D'Eugenio, F., Ji, X., Parlanti, E., Scholtz, J., Sun, F., Venturi, G., Bakx, T. J. L. C., Curti, M., Maiolino, R., Tacchella, S., Zavala, J. A., Hainline, K., Witstok, J., Johnson, B. D., Alberts, S., Bunker, A. J., Charlot, S., Eisenstein, D. J., Helton, J. M., Jakobsen, P., Kumari, N., Robertson, B., Saxena, A., Übler, H., Williams, C. C., Willmer, C. N. A., & Willott, C. 2024,a The eventful life of a luminous galaxy at $z = 14$: metal enrichment, feedback, and low gas fraction? *arXiv e-prints*, arXiv:2409.20533

Carniani, S., Hainline, K., D'Eugenio, F., Eisenstein, D. J., Jakobsen, P., Witstok, J., Johnson, B. D., Chevallard, J., Maiolino, R., Helton, J. M., Willott, C., Robertson, B., Alberts, S., Arribas, S., Baker, W. M., Bhatawdekar, R., Boyett, K., Bunker, A. J., Cameron, A. J., Cargile, P. A., Charlot, S., Curti, M., Curtis-Lake, E., Egami, E., Giardino, G., Isaak, K., Ji, Z., Jones, G. C., Kumari, N., Maseda, M. V., Parlanti, E., Pérez-González, P. G., Rawle, T., Rieke, G., Rieke, M., Del Pino, B. R., Saxena, A., Scholtz, J., Smit, R., Sun, F., Tacchella, S., Übler, H., Venturi, G., Williams, C. C., & Willmer, C. N. A. 2024,b Spectroscopic confirmation of two luminous galaxies at a redshift of 14. *Nature*, 633b(8029), 318–322

Casey, C. M., Akins, H. B., Shuntov, M., Ilbert, O., Paquereau, L., Franco, M., Hayward, C. C., Finkelstein, S. L., Boylan-Kolchin, M., Robertson, B. E., Allen, N., Brinch, M., Cooper, O. R., Ding, X., Drakos, N. E., Faisst, A. L., Fujimoto, S., Gillman, S., Harish, S., Hirschmann, M., Jin, S., Kartaltepe, J. S., Koekemoer, A. M., Kokorev, V., Liu, D., Long, A. S., Magdis, G., Maraston, C., Martin, C. L., McCracken, H. J., McKinney, J., Mobasher, B., Rhodes, J., Rich, R. M., Sanders, D. B., Silverman, J. D., Toft, S., Vijayan, A. P., Weaver, J. R., Wilkins, S. M., Yang, L., & Zavala, J. A. 2024, COSMOS-Web: Intrinsically Luminous $z \gtrsim 10$ Galaxy Candidates Test Early Stellar Mass Assembly. *ApJ*, 965(1), 98

Castellano, M., Fontana, A., Treu, T., Merlin, E., Santini, P., Bergamini, P., Grillo, C., Rosati, P., Acebron, A., Leethochawalit, N., Paris, D., Bonchi, A., Belfiori, D., Calabrò, A.,

Correnti, M., Nonino, M., Polenta, G., Trenti, M., Boyett, K., Brammer, G., Broadhurst, T., Caminha, G. B., Chen, W., Filippenko, A. V., Fortuni, F., Glazebrook, K., Mascia, S., Mason, C. A., Menci, N., Meneghetti, M., Mercurio, A., Metha, B., Morishita, T., Nanayakkara, T., Pentericci, L., Roberts-Borsani, G., Roy, N., Vanzella, E., Vulcani, B., Yang, L., & Wang, X. 2023, Early Results from GLASS-JWST. XIX. A High Density of Bright Galaxies at z $\approx$ 10 in the A2744 Region. *ApJ*, 948(2), L14

Curti, M., D'Eugenio, F., Carniani, S., Maiolino, R., Sandles, L., Witstok, J., Baker, W. M., Bennett, J. S., Piotrowska, J. M., Tacchella, S., Charlot, S., Nakajima, K., Maheson, G., Mannucci, F., Amiri, A., Arribas, S., Belfiore, F., Bonaventura, N. R., Bunker, A. J., Chevallard, J., Cresci, G., Curtis-Lake, E., Hayden-Pawson, C., Jones, G. C., Kumari, N., Laseter, I., Looser, T. J., Marconi, A., Maseda, M. V., Scholtz, J., Smit, R., Übler, H., & Wallace, I. E. B. 2023, The chemical enrichment in the early Universe as probed by JWST via direct metallicity measurements at z $\sim$ 8. *MNRAS*, 518(1), 425–438

de Graaff, A., Setton, D. J., Brammer, G., Cutler, S., Suess, K. A., Labbe, I., Leja, J., Weibel, A., Maseda, M. V., Whitaker, K. E., Bezanson, R., Boogaard, L. A., Cleri, N. J., De Lucia, G., Franx, M., Greene, J. E., Hirschmann, M., Matthee, J., McConachie, I., Naidu, R. P., Oesch, P. A., Price, S. H., Rix, H.-W., Valentino, F., Wang, B., & Williams, C. C. 2024, Efficient formation of a massive quiescent galaxy at redshift 4.9. *arXiv e-prints*, arXiv:2404.05683

Donnan, C. T., McLeod, D. J., Dunlop, J. S., McLure, R. J., Carnall, A. C., Begley, R., Cullen, F., Hamadouche, M. L., Bowler, R. A. A., Magee, D., McCracken, H. J., Milvang-Jensen, B., Moneti, A., & Targett, T. 2023, The evolution of the galaxy UV luminosity function at redshifts z $\sim 8-15$ from deep JWST and ground-based near-infrared imaging. *MNRAS*, 518(4), 6011–6040

Finkelstein, S. L., Bagley, M. B., Ferguson, H. C., Wilkins, S. M., Kartaltepe, J. S., Papovich, C., Yung, L. Y. A., Arrabal Haro, P., Behroozi, P., Dickinson, M., Kocevski, D. D., Koekemoer, A. M., Larson, R. L., Le Bail, A., Morales, A. M., Pérez-González, P. G., Burgarella, D., Davé, R., Hirschmann, M., Somerville, R. S., Wuyts, S., Bromm, V., Casey, C. M., Fontana, A., Fujimoto, S., Gardner, J. P., Giavalisco, M., Grazian, A., Grogin, N. A., Hathi, N. P., Hutchison, T. A., Jha, S. W., Jogee, S., Kewley, L. J., Kirkpatrick, A., Long, A. S., Lotz, J. M., Pentericci, L., Pierel, J. D. R., Pirzkal, N., Ravindranath, S., Ryan, R. E., Trump, J. R., Yang, G., Bhatawdekar, R., Bisigello, L., Buat, V., Calabrò, A., Castellano, M., Cleri, N. J., Cooper, M. C., Croton, D., Daddi, E., Dekel, A., Elbaz, D., Franco, M., Gawiser, E., Holwerda, B. W., Huertas-Company, M., Jaskot, A. E., Leung, G. C. K., Lucas, R. A., Mobasher, B., Pandya, V., Tacchella, S., Weiner, B. J., & Zavala, J. A. 2023, CEERS Key Paper. I. An Early Look into the First 500 Myr of Galaxy Formation with JWST. *ApJ*, 946(1), L13

Fudamoto, Y., Oesch, P. A., Schouws, S., Stefanon, M., Smit, R., Bouwens, R. J., Bowler, R. A. A., Endsley, R., Gonzalez, V., Inami, H., Labbe, I., Stark, D., Aravena, M., Barrufet, L., da Cunha, E., Dayal, P., Ferrara, A., Graziani, L., Hodge, J., Hutter, A., Li, Y., De Looze, I., Nanayakkara, T., Pallottini, A., Riechers, D., Schneider, R., Ucci, G., van der Werf, P., & White, C. 2021, Normal, dust-obscured galaxies in the epoch of reionization. *Nature*, 597(7877), 489–492

Glazebrook, K., Nanayakkara, T., Schreiber, C., Lagos, C., Kawinwanichakij, L., Jacobs, C., Chittenden, H., Brammer, G., Kacprzak, G. G., Labbe, I., Marchesini, D., Marsan, Z. C., Oesch, P. A., Papovich, C., Remus, R.-S., Tran, K.-V. H., Esdaile, J., & Chandro-Gomez, A. 2024, A massive galaxy that formed its stars at z $\approx$ 11. *Nature*, 628(8007), 277–281

Harikane, Y., Ouchi, M., Oguri, M., Ono, Y., Nakajima, K., Isobe, Y., Umeda, H., Mawatari, K., & Zhang, Y. 2023, A Comprehensive Study of Galaxies at z 9-16 Found in the Early JWST Data: Ultraviolet Luminosity Functions and Cosmic Star Formation History at the Pre-reionization Epoch. *ApJS*, 265(1), 5

Kennicutt, R. C. & Evans, N. J. 2012, Star Formation in the Milky Way and Nearby Galaxies. *ARA&A*, 50, 531–608

Labbé, I., van Dokkum, P., Nelson, E., Bezanson, R., Suess, K. A., Leja, J., Brammer, G., Whitaker, K., Mathews, E., Stefanon, M., & Wang, B. 2023, A population of red candidate massive galaxies 600 Myr after the Big Bang. *Nature*, 616(7956), 266–269

Mashian, N., Oesch, P. A., & Loeb, A. 2016, An empirical model for the galaxy luminosity and star formation rate function at high redshift. *MNRAS*, 455, 2101–2109

Mason, C. A., Trenti, M., & Treu, T. 2015, The Galaxy UV Luminosity Function before the Epoch of Reionization. *ApJ*, 813, 21

Naidu, R. P., Oesch, P. A., van Dokkum, P., Nelson, E. J., Suess, K. A., Brammer, G., Whitaker, K. E., Illingworth, G., Bouwens, R., Tacchella, S., Matthee, J., Allen, N., Bezanson, R., Conroy, C., Labbe, I., Leja, J., Leonova, E., Magee, D., Price, S. H., Setton, D. J., Strait, V., Stefanon, M., Toft, S., Weaver, J. R., & Weibel, A. 2022, Two Remarkably Luminous Galaxy Candidates at $z \approx$ 10-12 Revealed by JWST. *ApJ*, 940(1), L14

Nakajima, K., Ouchi, M., Isobe, Y., Harikane, Y., Zhang, Y., Ono, Y., Umeda, H., & Oguri, M. 2023, JWST Census for the Mass-Metallicity Star Formation Relations at z = 4-10 with Self-consistent Flux Calibration and Proper Metallicity Calibrators. *ApJS*, 269(2), 33

Nanayakkara, T., Glazebrook, K., Jacobs, C., Kawinwanichakij, L., Schreiber, C., Brammer, G., Esdaile, J., Kacprzak, G. G., Labbe, I., Lagos, C., Marchesini, D., Marsan, Z. C., Oesch, P. A., Papovich, C., Remus, R.-S., & Tran, K.-V. H. 2024, A population of faint, old, and massive quiescent galaxies at $3 < z < 4$ revealed by JWST NIRSpec Spectroscopy. *Scientific Reports*, 14, 3724

Oesch, P. A., Bouwens, R. J., Illingworth, G. D., Labbé, I., & Stefanon, M. 2018, The Dearth of $z \sim 10$ Galaxies in All HST Legacy Fields – The Rapid Evolution of the Galaxy Population in the First 500 Myr. *ApJ*, 855, 105

Rodighiero, G., Bisigello, L., Iani, E., Marasco, A., Grazian, A., Sinigaglia, F., Cassata, P., & Gruppioni, C. 2023, JWST unveils heavily obscured (active and passive) sources up to z 13. *MNRAS*, 518(1), L19–L24

Schouws, S., Bouwens, R. J., Ormerod, K., Smit, R., Algera, H., Sommovigo, L., Hodge, J., Ferrara, A., Oesch, P. A., Rowland, L. E., van Leeuwen, I., Stefanon, M., Herard-Demanche, T., Fudamoto, Y., Röttgering, H., & van der Werf, P. 2024, Detection of [OIII]88μm in JADES-GS-z14-0 at z=14.1793. *arXiv e-prints*, arXiv:2409.20549

Stefanon, M., Bouwens, R. J., Labbé, I., Illingworth, G. D., Gonzalez, V., & Oesch, P. A. 2021, Galaxy Stellar Mass Functions from z 10 to z 6 using the Deepest Spitzer/Infrared Array Camera Data: No Significant Evolution in the Stellar-to-halo Mass Ratio of Galaxies in the First Gigayear of Cosmic Time. *ApJ*, 922(1), 29

Trac, H., Cen, R., & Mansfield, P. 2015, SCORCH I: The Galaxy-Halo Connection in the First Billion Years. *ApJ*, 813, 54

Weibel, A., de Graaff, A., Setton, D. J., Miller, T. B., Oesch, P. A., Brammer, G., Lagos, C. D. P., Whitaker, K. E., Williams, C. C., Baggen, J. F. W., Bezanson, R., Boogaard, L. A., Cleri, N. J., Greene, J. E., Hirschmann, M., Hviding, R. E., Kuruvanthodi, A., Labbé, I., Leja, J., Maseda, M. V., Matthee, J., McConachie, I., Naidu, R. P., Roberts-Borsani, G., Schaerer, D., Suess, K. A., Valentino, F., van Dokkum, P., & Wang, B. 2024, RUBIES Reveals a Massive Quiescent Galaxy at z=7.3. *arXiv e-prints*, arXiv:2409.03829

Williams, C. C., Curtis-Lake, E., Hainline, K. N., Chevallard, J., Robertson, B. E., Charlot, S., Endsley, R., Stark, D. P., Willmer, C. N. A., Alberts, S., Amorin, R., Arribas, S., Baum, S., Bunker, A., Carniani, S., Crandall, S., Egami, E., Eisenstein, D. J., Ferruit, P., Husemann, B., Maseda, M. V., Maiolino, R., Rawle, T. D., Rieke, M., Smit, R., Tacchella, S., & Willott, C. J. 2018, The JWST Extragalactic Mock Catalog: Modeling Galaxy Populations from the UV through the Near-IR over 13 Billion Years of Cosmic History. *ApJS*, 236(2), 33

Xiao, M., Oesch, P., Elbaz, D., Bing, L., Nelson, E., Weibel, A., Illingworth, G., van Dokkum, P., Naidu, R., Daddi, E., Bouwens, R., Matthee, J., Wuyts, S., Chisholm, J., Brammer, G., Dickinson, M., Magnelli, B., Leroy, L., Schaerer, D., Herard-Demanche, T., Lim, S., Barrufet, L., Endsley, R., Fudamoto, Y., Gómez-Guijarro, C., Gottumukkala, R., Labbe, I., Magee, D., Marchesini, D., Maseda, M., Qin, Y., Reddy, N., Shapley, A., Shivaei, I., Shuntov, M., Stefanon, M., Whitaker, K., & Wyithe, J. S. 2023, Accelerated Formation of Ultra-Massive Galaxies in the First Billion Years. *arXiv e-prints*, arXiv:2309.02492

The First Chapters of Our Cosmic History with JWST
Proceedings IAU Symposium No. 391, 2025
T. Nanayakkara & M. Maseda, eds.
doi:10.1017/S1743921324002370

Stellar Populations and the ISM in the High-Redshift Universe as Revealed by JWST

Naveen A. Reddy

University of California, Riverside
email: naveen.reddy@ucr.edu

Abstract. In this contribution, I will review a few of the key characteristics of the stellar populations and interstellar medium (ISM) of high-redshift galaxies as revealed by James Webb Space Telescope (JWST) spectroscopy. Specifically, I will discuss recent evidence for nitrogen enhancement and proposed mechanisms to explain it, the existence of galaxies with very blue UV continuum slopes, and the ionization state of emission-line galaxies. I will then focus on some recent work to understand the connection between ionization parameter, gas density, and metallicity, showing that the gas density is an important factor in modulating the ionization parameter across a large range of redshift.

Keywords. galaxies: high-redshift; galaxies: ISM; galaxies: evolution; galaxies: star formation

---

1. Introduction

Within just three years of its launch, the James Webb Space Telescope (JWST) has transformed our understanding of the stellar populations and interstellar medium (ISM) of high-redshift galaxies. The first rest-frame UV through optical spectra obtained for $z \gtrsim 8$ galaxies with JWST exhibited line ratios suggestive of unique elemental abundances that appear to be rare at lower redshifts. On the other hand, rest-frame optical spectra of slightly lower-redshift galaxies (e.g., $z \sim 4-8$) have pointed to ISM physical conditions that continue a sequence of low-metallicity and high-ionization conditions suggested by pre-JWST ground-based spectroscopic data from multi-object near-IR spectroscopic surveys such as MOSDEF (Kriek et al. 2015). In this proceeding, I will review a few of the recent findings on the stellar populations and ISM of $z \gtrsim 2$, focusing on nitrogen enhancement, blue UV slopes, and the properties of emission-line galaxies. I will then focus on the factors that dictate the ionization state of the interstellar medium.

2. Nitrogen Enhancement

A number of recent JWST spectroscopic observations of galaxies up to $z \sim 11$ have uncovered a population of younger galaxies with elevated [N/O], indicating an excess of nitrogen beyond the value expected for standard chemical evolution models. The first such example is GN-z11 (Bunker et al. 2023; Figure 1). While the powering source of this galaxy is still debated (i.e., star formation versus AGN; Maiolino et al. (2024)), a number of other star-forming galaxies exhibiting N enhancement have now been discovered (Isobe et al. 2024; Marques-Chaves et al. 2024; Topping et al. 2024b). For the young stellar populations that characterize these galaxies, the primary enrichment of N occurs in the CNO cycle. In equilibrium, the CNO cycle produces more N relative to C and O atoms due to the slow reaction rate of N. The challenge for the high-redshift observations is

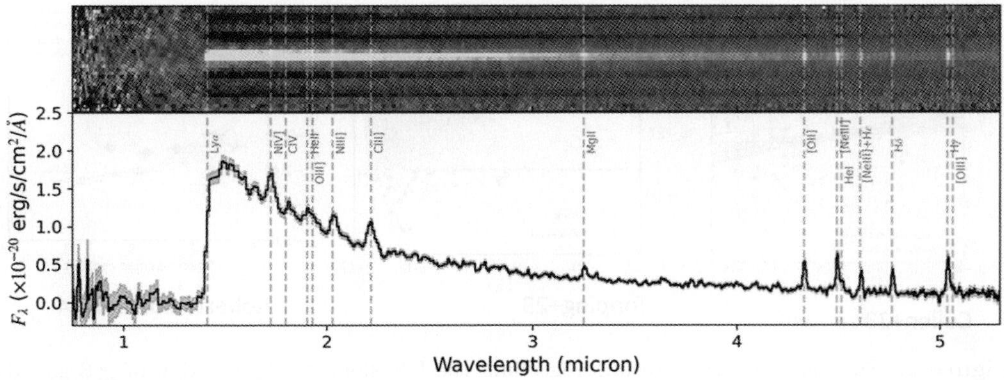

Figure 1. NIRSpec PRISM spectrum of GN-z11, at $z = 10.6$, showing strong NIV and NIII lines suggestive of a higher N abundance relative to O. **Taken from Bunker et al. (2023).**

to explain how this N-enriched material in the cores of massive stars (or in the shells of post main-sequence massive stars) is expelled into the ISM.

The steep temperature dependence of the CNO cycle, where the rate goes as T^{17}, implies a large temperature gradient resulting in convective transport. Subsequently, convective overshooting can efficiently bring N-enriched material from the interiors of massive stars closer to their surfaces. Additional mechanisms must be invoked to then expel this material into the ISM, almost all of which involve star formation occuring in dense environments. For example, tidal disruption events (Bunker et al. 2023; Cameron et al. 2023), dynamical encounters in dense star clusters (Cameron et al. 2023), strong winds from supermassive stars (Charbonnel et al. 2023), and wind eject from WR stars in dense environments (Senchyna et al. 2024) have all been invoked to explain the enhanced N abundance seen in some of the high-redshift galaxies. Such high-density environments are expected to be common given the high gas fractions and high star-formation-rate surface densities and compact sizes of high-redshift galaxies.

3. Blue UV Spectral Slopes

Another key finding of early JWST observations is the presence of galaxies with very blue UV spectral slopes. The UV slope, β, is sensitive to a variety of factors including the star-formation history, IMF, age, metallicity, and the strength of the nebular continuum. Dust attenuation causes the most significant changes in β, and thus the latter is often used as a proxy for the former (e.g., Meurer et al. 1999; Calzetti et al. 2000; Reddy et al. 2006, 2010, 2012; Daddi et al. 2007; Buat et al. 2012). JWST/NIRCam photometry of the rest-frame UV continuum emission of $z \gtrsim 8$ galaxies has revealed the presence of galaxies with blue UV slopes that are consistent with low-metallicity, low-dust, and young stellar populations (Cullen et al. 2023; Roberts-Borsani et al. 2024). Figure 2 highlights a few results from these studies, confirming the trends between UV slope and UV luminosity (or absolute magnitude) established at lower redshifts. For the most part, the existing observations suggest that current models are able to explain the range of observed UV slopes, and that "exotic" stellar populations and/or very top-heavy IMF are not needed to reproduce the observations.

4. Emission-Line Galaxies

JWST has not only revealed special classes of galaxies (e.g., those with enhanced N abundance, blue UV spectral slopes, as discussed above), but this facility has also given us insights into the more "typical" star-forming galaxies at $z \sim 3-9$. In particular, the

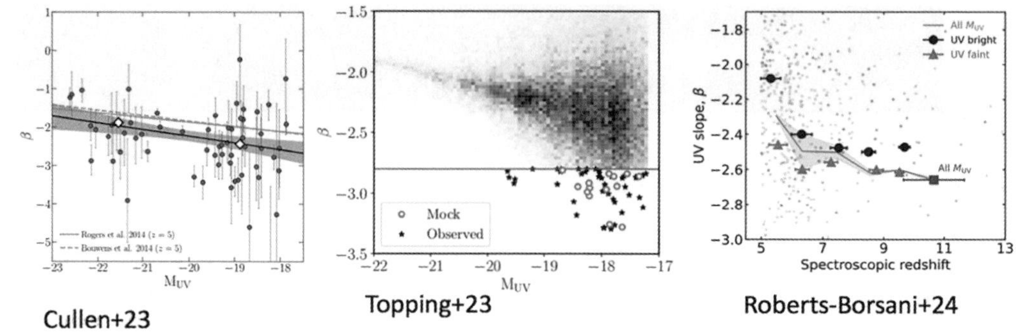

Figure 2. Summary of recent results regarding the UV slope (β distribution of $\gtrsim 8$ galaxies from JWST observations. The results indicate the presence of galaxies with blue UV slopes, many of which have slopes comparable to the bluer galaxies at lower redshifts. Some of the results indicate galaxies with extremely blue slopes that may be indicative of a young stellar population with a high escape fraction of ionizing photons. **Taken from Cullen et al. (2023); Topping et al. (2024a); Roberts-Borsani et al. (2024)**.

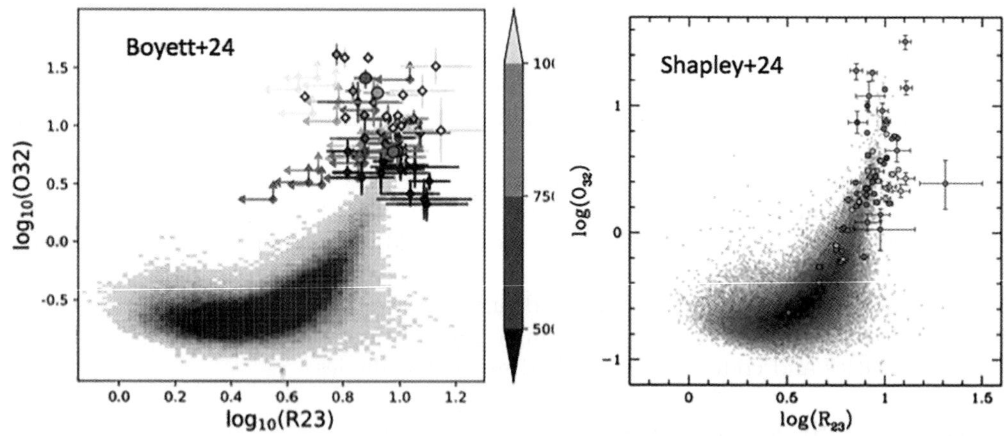

Figure 3. *Left:* O32 versus R23 (i.e., ionization parameter versus oxygen abundance) for $z \sim 3-9$ extreme emission line galaxies from Boyett et al. (2024), showng that these galaxies continue the local sequence towards high ionization and low metallicity. *Right:* Similar results from Shapley et al. (2024) that show that for typical star-forming galaxies at $z = 2.0 - 6.5$, the line ratios continue the same sequence in O32 and R23 seen in local star-forming galaxies. **Taken from Boyett et al. (2024); Shapley et al. (2024)**.

fraction of galaxies with high rest-frame optical emission line equivalent widths, the so-called extreme emission line galaxies (EELGs) appears to increase with redshift. Several studies, such as those highlighted in Figure 3, find line ratios for EELGs that occupy the same parameter space as lower redshift galaxies with high ionization and low metallicity (e.g., Boyett et al. 2024; Shapley et al. 2024).

5. Connection between Ionization Parameter and Gas Density

The ionization parameter (U)—commonly defined as the ratio of the number density of incident ionizing photons and the number density of hydrogen atoms—is a key property of the interstellar medium (ISM) that depends on the intensity and hardness of the ionizing radiation field, gas density, and the spatial distribution of gas relative to ionizing sources. Accordingly, variations in U can provide meaningful insights into how the

ISM and ionizing sources evolve with redshift and/or galaxy properties. Recent observations with JWST have revealed the presence of galaxies with apparently "extreme" ionization conditions at the epoch of reionization, where U may be an order of magnitude or more larger than that of typical star-forming galaxies at lower redshifts (Tang et al. 2023; Bunker et al. 2023). The strong inverse correlation between U and oxygen abundance (Z_{neb}) seen in the local Universe (Dopita and Evans 1986; Pérez-Montero 2014) is commonly interpreted as a reflection of the harder ionizing spectra associated with low metallicity stars. Consequently, the large values of U seen at high redshift are ostensibly due to the lower metallicity stellar populations characteristic of galaxies at these redshifts. While these results are intriguing, they also motivate a more detailed investigation of the physical causes of the elevated U inferred for high-redshift galaxies.

In this context, there have been a number of efforts focused on explaining the redshift evolution of U at a fixed stellar mass up to $z \sim 2$. Early work suggested that this evolution reflects the decrease in metallicity (and implied increase in hardness of the ionizing spectrum) of galaxies with redshift at a given stellar mass (e.g., Sanders et al. 2016; Kashino and Inoue 2019). On the other hand, because U is sensitive to the total ionizing photon rate (or SFR), the increase in SFR with redshift at a fixed stellar mass has also been invoked to explain in part the elevated U seen at high redshift (e.g., Kaasinen et al. 2017). Others have pointed to the increased gas (or SFR) surface densities and/or higher electron densities (n_e) as a possible factor in the redshift evolution of U (Brinchmann et al. 2008; Liu et al. 2008; Shimakawa et al. 2015; Masters et al. 2016; Bian et al. 2016; Davies et al. 2021; Papovich et al. 2022; Reddy et al. 2023a). Previous studies have necessarily been limited in the redshift range over which one can robustly constrain U and n_e, since the lines typically used to infer them for the same galaxies are inaccessible from the ground above $z \sim 4$.

JWST now allows access to the same diagnostics of U, Z_{neb}, and n_e for galaxies well into the epoch of reionization, where these properties are expected to lie outside the ranges typically observed up to $z \sim 2-3$, but which still overlap with the more "extreme" sources (i.e., with high U, low Z_{neb}, and high n_e) at these lower redshifts. Such measurements over larger dynamic ranges in redshift, stellar characteristics, and physical conditions in the ISM form an ideal dataset to evaluate the evolution in U. Here, I describe some of the efforts to use public JWST/NIRSpec spectroscopy obtained as part of the Cosmic Evolution Early Release Science (CEERS) survey to identify the factors responsible for modulating U in high-redshift ($z = 2.7 - 6.3$) galaxies.

5.1. *Electron Density and Ionizing Photon Rate*

For a radiation-bounded (spherical) nebula, the ionization parameter (U) is related to the ionizing photon rate (Q), hydrogen density (n_{H}), and Strömgren radius (R_{S}):

$$U \propto \frac{Q}{n_{\text{H}} R_{\text{S}}^2}. \tag{1}$$

The Strömgren radius is also related to Q and the volume-averaged hydrogen density, $\langle n_{\text{H}} \rangle$:

$$R_{\text{S}}^3 \propto \frac{Q}{\langle n_{\text{H}} \rangle^2} \propto \frac{Q}{\epsilon n_{\text{H}}^2}, \tag{2}$$

where the volume filling factor $\epsilon = (\langle n_{\text{H}} \rangle / n_{\text{H}})^2$. This factor represents the fraction of the HII region volume containing denser material that dominates the emission-line fluxes, such as [SII] (Osterbrock and Flather 1959; Kennicutt 1984). Combining Equations 1 and 2, and assuming all the gas is ionized (i.e., $n_{\text{H}} \sim n_e$, where n_e is the electron density),

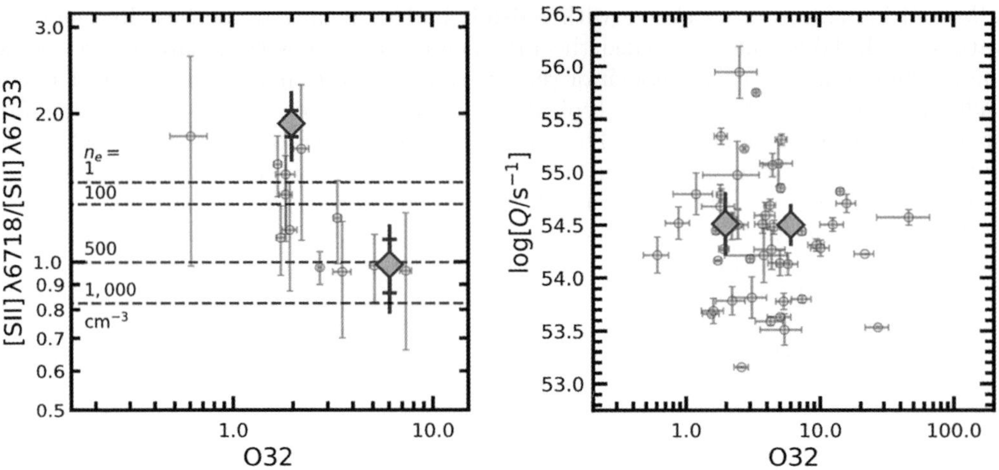

Figure 4. Variation of the [SII] doublet ratio (left) and Q (right) with O32. Individual galaxies with detections of [SII] and Q measurements are indicated by small circles in the top and bottom panels, respectively. Mean measurements from composite spectra of galaxies are shown by the diamonds. Measurement uncertainties for individual and composite values are indicated by the capped vertical lines. Estimates of sample variance (including measurement error) for the composite values are indicated by the uncapped vertical lines. In some cases, the error bars are smaller than the symbols. **Taken from Reddy et al. (2023b)**.

then yields the following dependence of U on Q, n_e, and ϵ:

$$U \propto [Q n_e \epsilon^2]^{1/3}. \tag{3}$$

Note that U is commonly defined as the ratio of the number density of ionizing photons over the number density of hydrogen atoms. For a radiation-bounded nebula, the number density of ionizing photons is simply proportional to the photon flux received at the surface of a sphere with radius R_S. An increase in density results in lower R_S and hence a higher photon flux. Thus, the overall dependence for a *radiation-bounded* nebula is one where U *increases* with $n_{\rm H}$ (or n_e). Motivated by the theoretical dependencies in Equation 3, we can examine how U (or its proxy, O32) varies with both Q and n_e. Using stacks of 24 galaxies at $z = 2.7 - 6.3$ in each of two bins of O32, Reddy et al. (2023b) showed that U is likely driven by changes in n_e rather than Q, where the former is estimated from the ratio of the [SII] doublet lines, and the latter is calculated from the dust-corrected Hα luminosity (see Figure 4).

5.2. *Dependence on Star-Formation-Rate Surface Density*

Previous efforts have established a connection between n_e and gas density, or SFR surface density (Shirazi et al. 2014; Shimakawa et al. 2015; Bian et al. 2016; Jiang et al. 2019; Davies et al. 2021; Reddy et al. 2023a). The *JWST* CEERS sample allows us to examine this connection at significantly higher redshifts ($z \gtrsim 2.7$) where galaxies may have a denser ISM on average. For this analysis, we considered the 22 galaxies in the sample that have existing measurements of effective half-light radii ($R_{\rm eff}$) from van der Wel et al. (2014). The SFR surface density is then defined as:

$$\Sigma_{\rm SFR} = \frac{{\rm SFR(H}\alpha)}{2\pi R_{\rm eff}^2}, \tag{4}$$

and is expressed in units of $M_\odot\,{\rm yr}^{-1}\,{\rm kpc}^{-2}$. The $\Sigma_{\rm SFR}$ subsample was divided into two equal-number bins of $\Sigma_{\rm SFR}$, and $\langle R_{[SII]}\rangle$ and $\langle {\rm O32}\rangle$ were computed from the composite

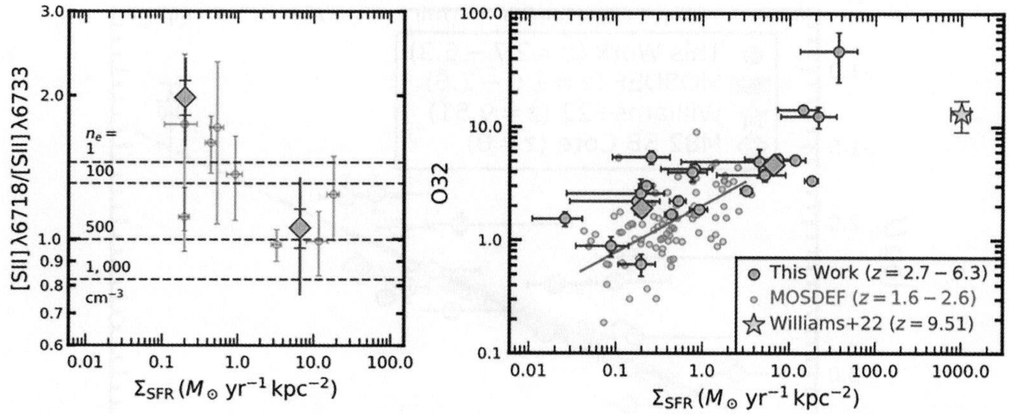

Figure 5. *Left:* Variation of $R_{[SII]}$ with $\Sigma_{\rm SFR}$. Individual galaxies with detections of [SII] are indicated by the small symbols. The average $\langle$[SII]$\rangle$ measured from the composite spectra are shown by the large diamonds. Capped vertical lines indicate measurement uncertainties, while uncapped vertical lines on the composite values include sample variance. *Right:* Variation of O32 with $\Sigma_{\rm SFR}$. Individual galaxies with robust $\Sigma_{\rm SFR}$ measurements are indicated by the blue circles, and the large diamonds indicate the average values computed from composite spectra in two bins of O32. Capped vertical lines indicate measurement uncertainties, while uncapped vertical lines on the composite values include sample variance. In some cases, the error bars are smaller than the symbols. Also shown are measurements from the MOSDEF survey (small red circles), along with the best-fit relation to those data (red line). For comparison, the $z = 9.51$ lensed galaxy from Williams et al. (2022) is indicated by the green star. **Taken from Reddy et al. (2023b).**

spectra constructed for galaxies in the two bins. The difference in $\langle R_{[SII]}\rangle$ between the low and high $\Sigma_{\rm SFR}$ bins is significant at the $\approx 4\sigma$ level, such that galaxies with higher $\Sigma_{\rm SFR}$ exhibit a lower $\langle R_{[SII]}\rangle$ (or higher $\langle n_e\rangle \simeq 500\,{\rm cm}^{-3}$), similar to the trend suggested by earlier studies (Figure 5).

Of the correlations investigated in this analysis, the one between O32 and $\Sigma_{\rm SFR}$ appears to be the most significant (Figure 5): a Spearman correlation test indicates a p value of $p = 1.97 \times 10^{-5}$. A similar and highly significant correlation was also found by Reddy et al. (2023a) for $z \sim 2 - 4$ galaxies in the MOSDEF survey. Data from that survey are also shown in Figure 5, along with the $z = 9.51$ lensed galaxy from Williams et al. (2022), for comparison.

Figure 6 directly shows the relationship between $\log U$ inferred from photoionization modeling and $\Sigma_{\rm SFR}$ for the 22 galaxies with measurements of the latter. Again, a Spearman correlation test implies a highly significant correlation between $\log U$ and $\Sigma_{\rm SFR}$, with $p = 3.2 \times 10^{-4}$. A formal linear fit to the sample of 22 galaxies yields the following relation between $\log U$ and $\log[\Sigma_{\rm SFR}/M_\odot\,{\rm yr}^{-1}\,{\rm kpc}^{-2}]$:

$$\log U = (0.378 \pm 0.021) \log \left[\frac{\Sigma_{\rm SFR}}{M_\odot\,{\rm yr}^{-1}\,{\rm kpc}^{-2}}\right] + (-2.655 \pm 0.011). \tag{5}$$

A similar fit to lower-redshift ($z \sim 1.9 - 3.7$) galaxies from the MOSDEF survey indicates a relation that is virtually identical to the one found here, despite the differences in the way the two samples were selected—i.e., rest-frame optical selection in the case of MOSDEF versus a heterogeneous selection in the case of CEERS. The starburst core of M82 (Förster Schreiber et al. 2001, 2003) is included in Figure 6, and is consistent with the trend between U and $\Sigma_{\rm SFR}$ for the high-redshift samples. For context, the $z = 9.51$ lensed galaxy in the RXJ2129 field reported by Williams et al. (2022) has $U = -1.24$ and an

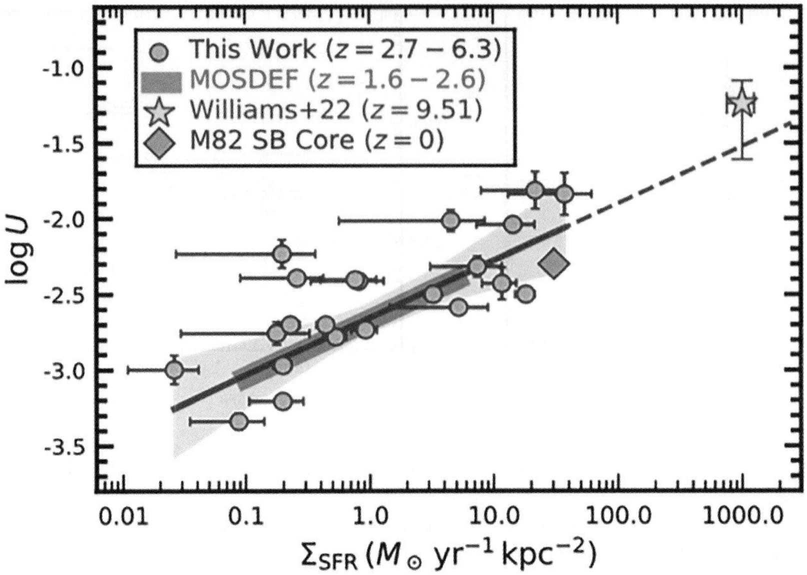

Figure 6. Variation of O32 with $\Sigma_{\rm SFR}$. Individual galaxies with robust $\Sigma_{\rm SFR}$ measurements are indicated by the blue circles, and the large diamonds indicate the average values computed from composite spectra in two bins of O32. Capped vertical lines indicate measurement uncertainties, while uncapped vertical lines on the composite values include sample variance. In some cases, the error bars are smaller than the symbols. Also shown are measurements from the MOSDEF survey (small red circles), along with the best-fit relation to those data (red line). For comparison, the $z = 9.51$ lensed galaxy from Williams et al. (2022) is indicated by the green star. **Taken from Reddy et al. (2023b)**.

extremely high $\Sigma_{\rm SFR} \simeq 1000\, M_\odot\, {\rm yr}^{-1}\, {\rm kpc}^{-2}$. These values place the $z \sim 9.5$ galaxy on the extrapolation of the U vs. $\Sigma_{\rm SFR}$ relation established from the lower-redshift samples. This agreement, along with the virtually identical U vs. $\Sigma_{\rm SFR}$ relations derived from the lower-redshift MOSDEF sample ($\langle z \rangle \simeq 2.1$) and the current sample ($\langle z \rangle \simeq 4.4$), may indicate that the U vs. $\Sigma_{\rm SFR}$ relation does not evolve with redshift. If additional measurements for high-$\Sigma_{\rm SFR}$ galaxies confirm this lack of evolution, it would imply that the processes that couple U to $\Sigma_{\rm SFR}$ are independent of redshift and that $\Sigma_{\rm SFR}$ is a key factor in modulating U at any redshift.

6. Open Questions

While JWST has given us important insights into different classes of objects at high redshift, the samples, particularly at $z \gtrsim 6$ are still rather small and we have yet to fully characterize the more typical star-forming galaxies at these redshifts. Upcoming larger spectroscopic surveys with JWST should illuminate the prevalence of galaxies with high N abundance, blue UV slopes, extreme emission line equivalent widths, and hard ionization states. Detailed photoionization modeling, particularly at lower redshifts ($z \sim 2-4$) where essentially all the important rest-frame optical emission lines are covered, will be essential for accurately constraining the nature of the ionizing sources and their effect on the ISM. Very deep spectra have already been used to measure weak auroral lines that can provide more precise constraints on multiple elemental abundances, including O, N, and S. Measurements of electron densities from, e.g., CIII], [SII], [ArIV], and HeI will be important for understanding the effect of massive stars on the ionization state of the ISM. Though not discussed here, JWST also promises a transformation in our understanding of the dust contents of high-redshift galaxies through the detection of multiple Balmer and/or Paschen emission lines, and measurements of the warm dust continuum emission and PAH emission with MIRI observations. When combined with ALMA constraints on cold dust emission, we should soon have a more comprehensive understanding of the dust contents and dust lifecycle at high redshift.

References

Bian, F., Kewley, L. J., Dopita, M. A., & Juneau, S. 2016, Local Analogs for High-redshift Galaxies: Resembling the Physical Conditions of the Interstellar Medium in High-redshift Galaxies. *Astrophysical Journal*, 822(2), 62

Boyett, K., Bunker, A. J., Curtis-Lake, E., Chevallard, J., Cameron, A. J., Jones, G. C., Saxena, A., Charlot, S., Curti, M., Wallace, I. E. B., Arribas, S., Carniani, S., Willott, C., Alberts, S., Eisenstein, D. J., Hainline, K., Hausen, R., Johnson, B. D., Rieke, M., Robertson, B., Stark, D. P., Tacchella, S., Williams, C. C., Chen, Z., Egami, E., Endsley, R., Kumari, N., Laseter, I., Looser, T. J., Maseda, M. V., Scholtz, J., Shivaei, I., Simmonds, C., Smit, R., Übler, H., & Witstok, J. 2024, Extreme emission line galaxies detected in JADES JWST/NIRSpec I: inferred galaxy properties. *Monthly Notices of the Royal Astronomical Society*

Brinchmann, J., Pettini, M., & Charlot, S. 2008, New insights into the stellar content and physical conditions of star-forming galaxies at z = 2-3 from spectral modelling. *Monthly Notices of the Royal Astronomical Society*, 385(2), 769–782

Buat, V., Noll, S., Burgarella, D., Giovannoli, E., Charmandaris, V., Pannella, M., Hwang, H. S., Elbaz, D., Dickinson, M., Magdis, G., Reddy, N., & Murphy, E. J. 2012, GOODS-Herschel: dust attenuation properties of UV selected high redshift galaxies. *Astronomy & Astrophysics*, 545, A141

Bunker, A. J., Saxena, A., Cameron, A. J., Willott, C. J., Curtis-Lake, E., Jakobsen, P., Carniani, S., Smit, R., Maiolino, R., Witstok, J., Curti, M., D'Eugenio, F., Jones, G. C., Ferruit, P., Arribas, S., Charlot, S., Chevallard, J., Giardino, G., de Graaff, A., Looser, T. J., Luetzgendorf, N., Maseda, M. V., Rawle, T., Rix, H.-W., Rodriguez Del Pino, B., Alberts, S., Egami, E., Eisenstein, D. J., Endsley, R., Hainline, K., Hausen, R., Johnson, B. D., Rieke, G., Rieke, M., Robertson, B. E., Shivaei, I., Stark, D. P., Sun, F., Tacchella, S., Tang, M., Williams, C. C., Willmer, C. N. A., Baker, W. M., Baum, S., Bhatawdekar, R., Bowler, R., Boyett, K., Chen, Z., Circosta, C., Helton, J. M., Ji, Z., Lyu, J., Nelson, E., Parlanti, E., Perna, M., Sandles, L., Scholtz, J., Suess, K. A., Topping, M. W., Uebler, H., Wallace, I. E. B., & Whitler, L. 2023, JADES NIRSpec Spectroscopy of GN-z11: Lyman-α emission and possible enhanced nitrogen abundance in a $z = 10.60$ luminous galaxy. *arXiv e-prints*, arXiv:2302.07256

Calzetti, D., Armus, L., Bohlin, R. C., Kinney, A. L., Koornneef, J., & Storchi-Bergmann, T. 2000, The Dust Content and Opacity of Actively Star-forming Galaxies. *Astrophysical Journal*, 533, 682–695

Cameron, A. J., Katz, H., Rey, M. P., & Saxena, A. 2023, Nitrogen enhancements 440 Myr after the big bang: supersolar N/O, a tidal disruption event, or a dense stellar cluster in GN-z11? *Monthly Notices of the Royal Astronomical Society*, 523(3), 3516–3525

Charbonnel, C., Schaerer, D., Prantzos, N., Ramírez-Galeano, L., Fragos, T., Kuruvanthodi, A., Marques-Chaves, R., & Gieles, M. 2023, N-enhancement in GN-z11: First evidence for supermassive stars nucleosynthesis in proto-globular clusters-like conditions at high redshift? *Astronomy & Astrophysics*, 673, L7

Cullen, F., McLure, R. J., McLeod, D. J., Dunlop, J. S., Donnan, C. T., Carnall, A. C., Bowler, R. A. A., Begley, R., Hamadouche, M. L., & Stanton, T. M. 2023, The ultraviolet continuum slopes (β) of galaxies at z = 8-16 from JWST and ground-based near-infrared imaging. *Monthly Notices of the Royal Astronomical Society*, 520(1), 14–23

Daddi, E., Dickinson, M., Morrison, G., Chary, R., Cimatti, A., Elbaz, D., Frayer, D., Renzini, A., Pope, A., Alexander, D. M., Bauer, F. E., Giavalisco, M., Huynh, M., Kurk, J., & Mignoli, M. 2007, Multiwavelength Study of Massive Galaxies at z˜2. I. Star Formation and Galaxy Growth. *Astrophysical Journal*, 670, 156–172

Davies, R. L., Schreiber, N. M. F., Genzel, R., Shimizu, T. T., Davies, R. I., Schruba, A., Tacconi, L. J., Übler, H., Wisnioski, E., Wuyts, S., Fossati, M., Herrera-Camus, R., Lutz, D., Mendel, J. T., Naab, T., Price, S. H., Renzini, A., Wilman, D., Beifiori, A., Belli, S., Burkert, A., Chan, J., Contursi, A., Fabricius, M., Lee, M. M., Saglia, R. P., & Sternberg, A. 2021, The KMOS^{3D} Survey: Investigating the Origin of the Elevated Electron Densities in Star-forming Galaxies at $1 \lesssim z \lesssim 3$. *Astrophysical Journal*, 909(1), 78

Dopita, M. A. & Evans, I. N. 1986, Theoretical models for H II regions. II - The extragalactic H II region abundance sequence. *Astrophysical Journal*, 307, 431–440

Förster Schreiber, N. M., Genzel, R., Lutz, D., Kunze, D., & Sternberg, A. 2001, Near-Infrared Integral Field Spectroscopy and Mid-Infrared Spectroscopy of the Starburst Galaxy M82. *Astrophysical Journal*, 552(2), 544–571

Förster Schreiber, N. M., Genzel, R., Lutz, D., & Sternberg, A. 2003, The Nature of Starburst Activity in M82. *Astrophysical Journal*, 599(1), 193–217

Isobe, Y., Ouchi, M., Tominaga, N., Watanabe, K., Nakajima, K., Umeda, H., Yajima, H., Harikane, Y., Fukushima, H., Xu, Y., Ono, Y., & Zhang, Y. Nitrogen-Rich, Dense, and Highly-Ionized Gas in z>6 Galaxies. In *EAS2024* 2024, 1433

Jiang, T., Malhotra, S., Yang, H., & Rhoads, J. E. 2019, Correlation between SFR Surface Density and Thermal Pressure of Ionized Gas in Local Analogs of High-redshift Galaxies. *Astrophysical Journal*, 872(2), 146

Kaasinen, M., Bian, F., Groves, B., Kewley, L. J., & Gupta, A. 2017, The COSMOS-[O II] survey: evolution of electron density with star formation rate. *Monthly Notices of the Royal Astronomical Society*, 465(3), 3220–3234

Kashino, D. & Inoue, A. K. 2019, Disentangling the physical parameters of gaseous nebulae and galaxies. *Monthly Notices of the Royal Astronomical Society*, 486(1), 1053–1069

Kennicutt, R. C., J. 1984, Structural properties of giant H II regions in nearby galaxies. *Astrophysical Journal*, 287, 116–130

Kriek, M., Shapley, A. E., Reddy, N. A., Siana, B., Coil, A. L., Mobasher, B., Freeman, W. R., de Groot, L., Price, S. H., Sanders, R., Shivaei, I., Brammer, G. B., Momcheva, I. G., Skelton, R. E., van Dokkum, P. G., Whitaker, K. E., Aird, J., Azadi, M., Kassis, M., Bullock, J. S., Conroy, C., Davé, R., Kereš, D., & Krumholz, M. 2015, The MOSFIRE Deep Evolution Field (MOSDEF) Survey: Rest-frame Optical Spectroscopy for ˜1500 H-selected Galaxies at $1.37 < z < 3.8$. *Astrophysical Journal Supplement*, 218, 15

Liu, X., Shapley, A. E., Coil, A. L., Brinchmann, J., & Ma, C.-P. 2008, Metallicities and Physical Conditions in Star-forming Galaxies at z ˜ 1.0-1.5. *Astrophysical Journal*, 678, 758–779

Maiolino, R., Scholtz, J., Witstok, J., Carniani, S., D'Eugenio, F., de Graaff, A., Übler, H., Tacchella, S., Curtis-Lake, E., Arribas, S., Bunker, A., Charlot, S., Chevallard, J., Curti, M., Looser, T. J., Maseda, M. V., Rawle, T. D., Rodríguez del Pino, B., Willott, C. J., Egami, E., Eisenstein, D. J., Hainline, K. N., Robertson, B., Williams, C. C., Willmer, C. N. A., Baker, W. M., Boyett, K., DeCoursey, C., Fabian, A. C., Helton, J. M., Ji, Z., Jones, G. C., Kumari, N., Laporte, N., Nelson, E. J., Perna, M., Sandles, L., Shivaei, I., & Sun, F. 2024, A small and vigorous black hole in the early Universe. *Nature*, 627(8002), 59–63

Marques-Chaves, R., Schaerer, D., Kuruvanthodi, A., Korber, D., Prantzos, N., Charbonnel, C., Weibel, A., Izotov, Y. I., Messa, M., Brammer, G., Dessauges-Zavadsky, M., & Oesch, P. 2024, Extreme N-emitters at high redshift: Possible signatures of supermassive stars and globular cluster or black hole formation in action. *Astronomy & Astrophysics*, 681, A30

Masters, D., Faisst, A., & Capak, P. 2016, A Tight Relation between N/O Ratio and Galaxy Stellar Mass Can Explain the Evolution of Strong Emission Line Ratios with Redshift. *Astrophysical Journal*, 828(1), 18

Meurer, G. R., Heckman, T. M., & Calzetti, D. 1999, Dust Absorption and the Ultraviolet Luminosity Density at z˜3 as Calibrated by Local Starburst Galaxies. *Astrophysical Journal*, 521, 64–80

Osterbrock, D. & Flather, E. 1959, Electron Densities in the Orion Nebula. II. *Astrophysical Journal*, 129, 26

Papovich, C., Simons, R. C., Estrada-Carpenter, V., Matharu, J., Momcheva, I., Trump, J. R., Backhaus, B. E., Brammer, G., Cleri, N. J., Finkelstein, S. L., Giavalisco, M., Ji, Z., Jung, I., Kewley, L. J., Nicholls, D. C., Pirzkal, N., Rafelski, M., & Weiner, B. 2022, CLEAR: The Ionization and Chemical-enrichment Properties of Galaxies at 1.1 < z < 2.3. *Astrophysical Journal*, 937(1), 22

Pérez-Montero, E. 2014, Deriving model-based T_e-consistent chemical abundances in ionized gaseous nebulae. *Monthly Notices of the Royal Astronomical Society*, 441, 2663–2675

Reddy, N., Dickinson, M., Elbaz, D., Morrison, G., Giavalisco, M., Ivison, R., Papovich, C., Scott, D., Buat, V., Burgarella, D., Charmandaris, V., Daddi, E., Magdis, G., Murphy, E., Altieri, B., Aussel, H., Dannerbauer, H., Dasyra, K., Hwang, H. S., Kartaltepe, J., Leiton, R., Magnelli, B., & Popesso, P. 2012, GOODS-Herschel Measurements of the Dust Attenuation of Typical Star-forming Galaxies at High Redshift: Observations of Ultraviolet-selected Galaxies at z ˜ 2. *Astrophysical Journal*, 744, 154

Reddy, N. A., Erb, D. K., Pettini, M., Steidel, C. C., & Shapley, A. E. 2010, Dust Obscuration and Metallicity at High Redshift: New Inferences from UV, Hα, and 8 μm Observations of z ˜ 2 Star-forming Galaxies. *Astrophysical Journal*, 712, 1070–1091

Reddy, N. A., Sanders, R. L., Shapley, A. E., Topping, M. W., Kriek, M., Coil, A. L., Mobasher, B., Siana, B., & Rezaee, S. 2023,a The Impact of Star-Formation-Rate Surface Density on the Electron Density and Ionization Parameter of High-Redshift Galaxies. *arXiv e-prints*, arXiv:2302.10213.

Reddy, N. A., Steidel, C. C., Erb, D. K., Shapley, A. E., & Pettini, M. 2006, A Spectroscopic Survey of Redshift 1.4<˜z<˜3.0 Galaxies in the GOODS-North Field: Survey Description, Catalogs, and Properties. *Astrophysical Journal*, 653, 1004–1026

Reddy, N. A., Topping, M. W., Sanders, R. L., Shapley, A. E., & Brammer, G. 2023,b A JWST/NIRSpec Exploration of the Connection between Ionization Parameter, Electron Density, and Star-formation-rate Surface Density in z = 2.7-6.3 Galaxies. *Astrophysical Journal*, 952b(2), 167

Roberts-Borsani, G., Treu, T., Shapley, A., Fontana, A., Pentericci, L., Castellano, M., Morishita, T., Bergamini, P., & Rosati, P. 2024, Between the Extremes: A JWST Spectroscopic Benchmark for High Redshift Galaxies Using ˜500 Confirmed Sources at $z \gtrsim 5$. *arXiv e-prints*, arXiv:2403.07103

Sanders, R. L., Shapley, A. E., Kriek, M., Reddy, N. A., Freeman, W. R., Coil, A. L., Siana, B., Mobasher, B., Shivaei, I., Price, S. H., & de Groot, L. 2016, The MOSDEF Survey: Electron Density and Ionization Parameter at z ˜ 2.3. *Astrophysical Journal*, 816, 23

Senchyna, P., Plat, A., Stark, D. P., Rudie, G. C., Berg, D., Charlot, S., James, B. L., & Mingozzi, M. 2024, GN-z11 in Context: Possible Signatures of Globular Cluster Precursors at Redshift 10. *Astrophysical Journal*, 966(1), 92

Shapley, A. E., Sanders, R. L., Topping, M. W., Reddy, N. A., Berg, D. A., Bouwens, R. J., Brammer, G., Carnall, A. C., Cullen, F., Davé, R., Dunlop, J. S., Ellis, R. S., Förster Schreiber, N. M., . Furlanetto, S. R., Glazebrook, K., Illingworth, G. D., Jones, T., Kriek, M., McLeod, D. J., McLure, R. J., Narayanan, D., Oesch, P., Pahl, A. J., Pettini, M., Schaerer, D., Stark, D. P., Steidel, C. C., Tang, M., Clarke, L., Donnan, C. T., & Kehoe, E. 2024, The AURORA Survey: A New Era of Emission-line Diagrams with JWST/NIRSpec. *arXiv e-prints*, arXiv:2407.00157

Shimakawa, R., Kodama, T., Steidel, C. C., Tadaki, K.-i., Tanaka, I., Strom, A. L., Hayashi, M., Koyama, Y., Suzuki, T. L., & Yamamoto, M. 2015, Correlation between star formation activity and electron density of ionized gas at z = 2.5. *Monthly Notices of the Royal Astronomical Society*, 451(2), 1284–1289

Shirazi, M., Brinchmann, J., & Rahmati, A. 2014, Stars were Born in Significantly Denser Regions in the Early Universe. *Astrophysical Journal*, 787(2), 120

Tang, M., Stark, D. P., Chen, Z., Mason, C., Topping, M., Endsley, R., Senchyna, P., Plat, A., Lu, T.-Y., Whitler, L., Robertson, B., & Charlot, S. 2023, JWST/NIRSpec Spectroscopy of $z = 7 - 9$ Star Forming Galaxies with CEERS: New Insight into Bright Lyα Emitters in Ionized Bubbles. *arXiv e-prints*, arXiv:2301.07072

Topping, M. W., Stark, D. P., Endsley, R., Whitler, L., Hainline, K., Johnson, B. D., Robertson, B., Tacchella, S., Chen, Z., Alberts, S., Baker, W. M., Bunker, A. J., Carniani, S., Charlot, S., Chevallard, J., Curtis-Lake, E., DeCoursey, C., Egami, E., Eisenstein, D. J., Ji, Z., Maiolino, R., Williams, C. C., Willmer, C. N. A., Willott, C., & Witstok, J. 2024,a The UV continuum slopes of early star-forming galaxies in JADES. *Monthly Notices of the Royal Astronomical Society*, 529a(4), 4087–4103

Topping, M. W., Stark, D. P., Senchyna, P., Chen, Z., Zitrin, A., Endsley, R., Charlot, S., Furtak, L. J., Maseda, M. V., Plat, A., Smit, R., Mainali, R., Chevallard, J., Molyneux, S., & Rigby, J. R. 2024,b Deep rest-UV JWST/NIRSpec spectroscopy of early galaxies: the demographics of CIV and N-emitters in the reionization era. *arXiv e-prints*, arXiv:2407.19009

van der Wel, A., Chang, Y.-Y., Bell, E. F., Holden, B. P., Ferguson, H. C., Giavalisco, M., Rix, H. W., Skelton, R., Whitaker, K., Momcheva, I., Brammer, G., Kassin, S. A., Martig, M., Dekel, A., Ceverino, D., Koo, D. C., Mozena, M., van Dokkum, P. G., Franx, M., Faber, S. M., & Primack, J. 2014, Geometry of Star-forming Galaxies from SDSS, 3D-HST, and CANDELS. *Astrophysical Journal Letters*, 792(1), L6

Williams, H., Kelly, P. L., Chen, W., Brammer, G., Zitrin, A., Treu, T., Scarlata, C., Koekemoer, A. M., Oguri, M., Lin, Y.-H., Diego, J. M., Nonino, M., Hjorth, J., Langeroodi, D., Broadhurst, T., Rogers, N., Perez-Fournon, I., Foley, R. J., Jha, S., Filippenko, A. V., Strolger, L., Pierel, J., Poidevin, F., & Yang, L. 2022, A Highly Magnified and Extremely Compact Galaxy at Redshift 9.51 with Strong Nebular Emission. *arXiv e-prints*, arXiv:2210.15699

The First Chapters of Our Cosmic History with JWST
Proceedings IAU Symposium No. 391, 2025
T. Nanayakkara & M. Maseda, eds.
doi:10.1017/S1743921324002436

Where is the floor? Studying the lowest metallicity galaxies with NIRSpec

Michael V. Maseda

Department of Astronomy, University of Wisconsin-Madison, 475 N. Charter St., Madison, WI 53706 USA
email: maseda@astro.wisc.edu

Abstract. Directly observing "Pop III" or zero-metallicity galaxies has long been a goal in our quest to find the first galaxies that formed in the Universe. These objects, however, have remained elusive even with 2.5 years of JWST observations. We present spectroscopy on the lowest metallicity systems yet confirmed with JWST, including the most extreme Lyman-alpha emitters at $M_{UV} = -15$ ($0.01\ L^\star$) detected with MUSE at $z = 3 - 7$ as well as a photometrically-selected sample at $z = 4 - 9$ from the JADES survey. While these objects show significantly lower metallicities than more UV-luminous objects at the same redshifts, we find a consistent metallicity floor at $\approx 2\%\ Z_\odot$. This indicates rapid chemical enrichment must be taking place in systems with light-weighted ages less than 5 Myr. However, many typical selections involving photometric redshifts and/or UV continuum slopes are biased against identification of even lower metallicities. Therefore, it is currently not possible to establish if this "floor" is real or an observational artifact. Spectroscopy of larger samples selected in less restrictive ways will be required to fully determine the answer.

Keywords. galaxies: abundances, galaxies: formation, galaxies: high-redshift

1. Introduction

In the local Universe, there is an observed metallicity floor around $12 + \log(O/H) = 6.9 - 7.1$ (i.e. $\approx 2\%\ Z_\odot$ Izotov et al. 2006; Izotov, Thuan, & Guseva 2012). The origin of this floor is not well understood, specifically since it is one to two orders of magnitude larger than the expected metallicity for the transition from "primordial" Population III to Population II stars (e.g. Wise et al. 2012). One potential origin has to do with the unique star formation histories that must be required for an extremely low-metallicity galaxy to exist at late cosmic times: Bortolini et al. (2024) study I Zw 18 with JWST/NIRCam, finding star formation histories that vary across the galaxy with some regions containing old (> 1 Gyr) stars even though the galaxy is currently undergoing its most prominent episode of star formation. Such a combination, with some extremely young and some much older stellar populations, is seemingly common in the local galaxies that constitute the "floor" in metallicity (Legrand et al. 2001). This, however, is impossible for galaxies at a redshift $\gtrsim 6$ where the Universe was less than 1 Gyr old, making comparisons between the physical conditions in local extremely metal poor galaxies and the highest-redshift galaxies in the Universe limited in their utility.

The advent of JWST (Gardner et al. 2023) and the NIRSpec spectrograph (Ferruit et al. 2022; Jakobsen et al. 2022) have facilitated observations of galaxies beyond $z = 6$ consistently for the first time. Measurements of gas-phase elemental abundances, namely Oxygen, are now commonplace and new calibrations have been developed to accurately determine them using e.g. observed ratios of strong emission lines (Laseter et al.

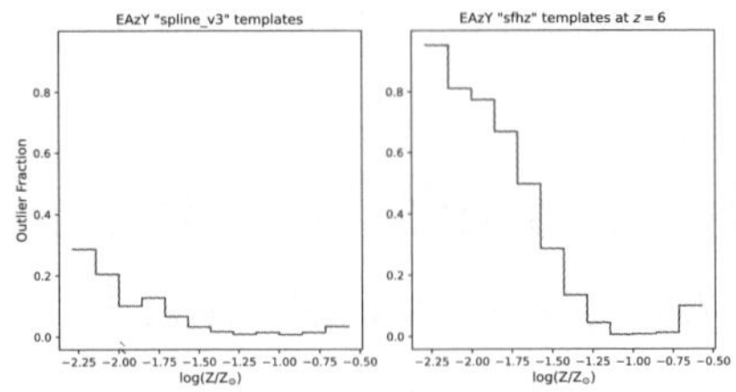

Figure 1. Catastrophic outlier fraction ($|\Delta z|/(1+z) > 0.5$) as a function of galaxy metallicity for sources at $z \approx 6$ (derived from the `JAGUAR` mock catalog; Williams et al. 2018) using the "sfhz" and "spline_v3" template sets in `EAzY`. `EAzY` is run on the observed JWST/NIRCam photometry at a depth consistent with the "medium" tier of JADES (Eisenstein et al. 2023). There is a clear trend regardless of the template choice where galaxies with lower metallicities are much more likely to be fit at the wrong redshift, typically $z \approx 0$. This has implications for photometric pre-selections and subsequent spectroscopic follow-up with NIRSpec.

2024). However, even after more than two years of science operations and hundreds of $z \gtrsim 6$ spectra, we have yet to directly observe a galaxy with a lower Oxygen abundance than the lowest-metallicity galaxies in the local Universe. The only exception to this is a strongly-lensed star cluster from Vanzella et al. (2023), but this metallicity measurement is not averaged over an entire galaxy.

Since NIRSpec requires a photometric pre-selection, the larger question becomes: are we capable of selecting galaxies for spectroscopy that are lower metallicity than $\approx 2\%\ Z_\odot$?

2. Observations and Results

When using a consistent set of high-z appropriate metallicity calibrations, e.g. those from (Laseter et al. 2024), the lowest-metallicity galaxies at $z \gtrsim 4$ are still at or above $\approx 2\%\ Z_\odot$ (Maseda et al. 2023; Atek et al. 2024; Chemerynska et al. 2024; Curti et al. 2024). The implied ages based on spectral fitting and line equivalent widths are extremely young, typically $\lesssim 5$ Myr. It remains a question, however, if this is a fundamental limit to the timescale for chemical enrichment in the early Universe or if we are simply not observing galaxies that are even younger and/or more metal-poor.

The galaxies from the aforementioned surveys were predominantly selected for NIRSpec observation on the basis of some kind of photometric redshift estimate (either observed colors or a photometric fitting code like `EAzY`; Brammer, van Dokkum, & Coppi 2008), with a sub-prioritization based on subjective criteria. Often, galaxies with steep UV continuum slopes ($\beta \lesssim -2$) are considered to be star-forming, as well as galaxies with strong optical emission lines which can be luminous enough to be detected in the photometry. All of these implicit requirements, however, can bias against the selection of galaxies with even lower metallicities.

2.1. *Photometric Redshifts*

As described above, all NIRSpec observations are predicated on a photometric pre-selection. This selection is typically done with HST/JWST photometry and, frequently, a photometric redshift code such as `EAzY` (Brammer, van Dokkum, & Coppi 2008). There are several assumptions underpinning the photometric redshift estimates, most prominently are the choice of template sets and a redshift/magnitude prior. We focus on the former assumption here, showing in Figure 1 that (1) the choice of template set has a strong impact on the total outlier fraction and (2) most template sets are systematically biased to fail more frequently on galaxies with metallicities $\lesssim 10\%\ Z_\odot$.

Specifically, the failure at low metallicities can be attributed to two main causes. The primary cause is the ratio of emission line fluxes as a function of metallicity: the ratio of

[OIII] to Hβ increases with decreasing metallicity down to $\approx 10-20\%$ Z$_\odot$, before declining again at even lower metallicities (e.g. Maseda et al. 2023). Since many of the very lowest metallicity galaxies have very large specific star formation rates, these emission lines have very large equivalent widths and can strongly affect the observed broad-band photometry (Maseda et al. 2014). Therefore, if the templates do not include galaxies with such low metallicities, even small changes in the line ratios can result in poor quality-of-fit. For example, a galaxy with Z = 1% Z$_\odot$ has 4 times less [OIII] luminosity than a Z = 10% Z$_\odot$ galaxy at the same star formation rate. This would result in a lower combined equivalent width for the two lines, which increases the χ^2 value when comparing to templates that only include sources with Z $\gtrsim$ 10% Z$_\odot$. Most frequently, at $z > 6$ `EAzY` will determine that the photometric redshift estimate is more likely to be around $z = 0$ as the HST photometry tracing the region at or below the real Lyman break is significantly shallower than the NIRCam photometry which covers the rest-optical emission lines. By virtue of minimizing χ^2, the very lowest metallicity galaxies are systematically put at the wrong redshift and would therefore never make it into samples to be observed with NIRSpec.

2.2. *UV Continuum Slopes*

In observations based on data from HST, trends and extrapolations suggest that galaxies at both higher redshift and lower UV luminosities should have progressively steeper UV continuum slopes (Bouwens et al. 2012). These results have been largely confirmed and extended with JWST (Cullen et al. 2024), where the highest redshifts suggest a population approaching the theoretical bluest-possible value of $\beta = -2.6$ on average. However, the impact of nebular continuum emission on the observed β values is such that at the very youngest ages and lowest metallicities, galaxies become redder (Raiter, Schaerer, & Fosbury 2010; Maseda et al. 2020; Katz et al. 2024). The very lowest metallicity galaxies, then, are indistinguishable from more metal-rich galaxies with significant amounts of dust attenuation. This means that prioritizing the bluest UV slopes, either in spectroscopic prioritization or in traditional photometric color cuts involving the Lyman break and rest-UV colors, will result in "purer" samples of low-metallicity star forming galaxies but will systematically miss the very lowest-metallicity and youngest galaxies.

3. Discussion

While JWST has given us the opportunity to observe galaxies at higher redshifts than previously possible, observing galaxies forming stars for the very first time is limited by our ability to identify these systems in the first place. Systematic biases due to templates (and priors) in photometric redshift fitting codes and the degeneracy between an extremely metal-poor galaxy and a metal-rich dusty galaxy make photometric preselections fraught. The way to make advances, then, is to either obtain much larger spectroscopic samples with relaxed selection criteria (such as allowing for redder β slopes) and larger contamination or to select galaxies in a very different way. In the latter case, we could perhaps rely on large-area slitless grism spectroscopic programs that do not require a pre-selection at all. This would require a significant investment of observing time to obtain data of sufficient depth and wavelength coverage. In either case, a large investment of time is necessary to establish whether or not we can truly find a galaxy made up entirely of "Pop III" stars which would only exist for a fleeting time.

References

Atek H., Labbé I., Furtak L. J., Chemerynska I., Fujimoto S., Setton D. J., Miller T. B., et al., 2024, *Nature*, 626, 975. doi:10.1038/s41586-024-07043-6

Bortolini G., Östlin G., Habel N., Hirschauer A. S., Jones O. C., Justtanont K., Meixner M., et al., 2024, arXiv, arXiv:2406.17429. doi:10.48550/arXiv.2406.17429

Bouwens R. J., Illingworth G. D., Oesch P. A., Franx M., Labbé I., Trenti M., van Dokkum P., et al., 2012, *ApJ*, 754, 83. doi:10.1088/0004-637X/754/2/83

Brammer G. B., van Dokkum P. G., Coppi P., 2008, *ApJ*, 686, 1503. doi:10.1086/591786

Chemerynska I., Atek H., Dayal P., Furtak L. J., Feldmann R., Greene J. E., Maseda M. V., et al., 2024, arXiv, arXiv:2407.17110. doi:10.48550/arXiv.2407.17110

Cullen F., McLeod D. J., McLure R. J., Dunlop J. S., Donnan C. T., Carnall A. C., Keating L. C., et al., 2024, *MNRAS*, 531, 997. doi:10.1093/mnras/stae1211

Curti M., Maiolino R., Curtis-Lake E., Chevallard J., Carniani S., D'Eugenio F., Looser T. J., et al., 2024, *A&A*, 684, A75. doi:10.1051/0004-6361/202346698

Eisenstein D. J., Willott C., Alberts S., Arribas S., Bonaventura N., Bunker A. J., Cameron A. J., et al., 2023, arXiv, arXiv:2306.02465. doi:10.48550/arXiv.2306.02465

Ferruit P., Jakobsen P., Giardino G., Rawle T., Alves de Oliveira C., Arribas S., Beck T. L., et al., 2022, *A&A*, 661, A81. doi:10.1051/0004-6361/202142673

Gardner J. P., Mather J. C., Abbott R., Abell J. S., Abernathy M., Abney F. E., Abraham J. G., et al., 2023, *PASP*, 135, 068001. doi:10.1088/1538-3873/acd1b5

Izotov Y. I., Stasińska G., Meynet G., Guseva N. G., Thuan T. X., 2006, *A&A*, 448, 955. doi:10.1051/0004-6361:20053763

Izotov Y. I., Thuan T. X., Guseva N. G., 2012, *A&A*, 546, A122. doi:10.1051/0004-6361/201219733

Jakobsen P., Ferruit P., Alves de Oliveira C., Arribas S., Bagnasco G., Barho R., Beck T. L., et al., 2022, *A&A*, 661, A80. doi:10.1051/0004-6361/202142663

Katz H., Cameron A. J., Saxena A., Barrufet L., Choustikov N., Cleri N. J., de Graaff A., et al., 2024, arXiv, arXiv:2408.03189. doi:10.48550/arXiv.2408.03189

Laseter I. H., Maseda M. V., Curti M., Maiolino R., D'Eugenio F., Cameron A. J., Looser T. J., et al., 2024, *A&A*, 681, A70. doi:10.1051/0004-6361/202347133

Legrand F., Tenorio-Tagle G., Silich S., Kunth D., Cerviño M., 2001, *ApJ*, 560, 630. doi:10.1086/322960

Maseda M. V., van der Wel A., Rix H.-W., da Cunha E., Pacifici C., Momcheva I., Brammer G. B., et al., 2014, *ApJ*, 791, 17. doi:10.1088/0004-637X/791/1/17

Maseda M. V., Bacon R., Lam D., Matthee J., Brinchmann J., Schaye J., Labbe I., et al., 2020, *MNRAS*, 493, 5120. doi:10.1093/mnras/staa622

Maseda M. V., Lewis Z., Matthee J., Hennawi J. F., Boogaard L., Feltre A., Nanayakkara T., et al., 2023, *ApJ*, 956, 11. doi:10.3847/1538-4357/acf12b

Raiter A., Schaerer D., Fosbury R. A. E., 2010, *A&A*, 523, A64. doi:10.1051/0004-6361/201015236

Vanzella E., Loiacono F., Bergamini P., Meštrić U., Castellano M., Rosati P., Meneghetti M., et al., 2023, *A&A*, 678, A173. doi:10.1051/0004-6361/202346981

Williams C. C., Curtis-Lake E., Hainline K. N., Chevallard J., Robertson B. E., Charlot S., Endsley R., et al., 2018, *ApJS*, 236, 33. doi:10.3847/1538-4365/aabcbb

Wise J. H., Turk M. J., Norman M. L., Abel T., 2012, *ApJ*, 745, 50. doi:10.1088/0004-637X/745/1/50

The First Chapters of Our Cosmic History with JWST
Proceedings IAU Symposium No. 391, 2025
T. Nanayakkara & M. Maseda, eds.
doi:10.1017/S1743921324002382

The origin of unusual chemical abundances from the JWST

Dyna Ibrahim and Chiaki Kobayashi
Centre for Astrophysics Research, University of Hertfordshire
email: d.ibrahim3@herts.ac.uk

Abstract. To understand the formation and evolution of the Universe, it is crucial to understand how and when the first stars formed. The latest observational data reveal unprecedented information about the chemical enrichment of the early Universe, which seems to behave differently from the local Universe. The first stars, being very massive, enrich their metal-poor environment in an uncertain way. In order to predict the abundances of the first galaxies, we include nucleosynthesis yields from Population III stars up to $300\mathrm{M}_\odot$, including faint supernovae, Wolf-Rayet (WR) stars and Pair-Instability Supernovae (PISN) into our state-of-the-art hydrodynamical cosmological simulations. Our code (based on Gadget-3) also includes the latest nucleosynthesis yields from population II stars (from Kobayashi et al. 2020) for all stellar mass ranges. We predict the chemical abundance evolution of galaxies for different elements from the early Universe to the local Universe. We first test the modelling of stellar feedback by comparing the observed evolution of mass–metallicity relations (MZR) and metallicity gradients of the interstellar medium. We then compare our model including Population III stars with observational data from the James Web Space Telescope (JWST). For elemental abundances, we find that the N/O abundance gives a systematically higher value, comparable to the observational data of very high-redshift galaxies such as GN-z11.

Keywords. galaxies: abundances – galaxies: formation – galaxies: evolution – methods: numerical.

1. Introduction

Recent observational data from the James Webb Space Telescope (JWST) pushed the barriers to our understanding of galaxy formation and evolution in the early Universe. These discoveries have shown surprisingly strong emission lines in very high-redshift galaxies, for instance, the Nitrogen-to-Oxygen abundance ratios of the galaxy GN-z11 at $z = 10.6$ (Bunker et al. 2023). Other galaxies have then been observed to show similar features (e.g., Marques-Chaves et al. 2024, $z = 8.68$; Isobe et al. 2023, $z = 6.23$; Topping et al. 2024a, $z = 6.1$; Topping et al. 2024b, $z = 7.04$; Castellano et al. 2024, $z = 12.34$; Schaerer et al. 2024, $z = 9.3$).

There are already various scenarios trying to explain the observed high N/O ratio. For example, using a correlation with globular clusters (e.g. Charbonnel et al. 2023; Senchyna et al. 2024), while others suggest that the enrichment comes from supermassive stars (Nagele and Umeda 2023; Nandal et al. 2024). In the very early Universe, the properties of the first stars may be very different. This includes the stellar mass function (which can be top-heavy) and the nucleosynthesis yields. Another hypothesis is from Kobayashi and Ferrara (2024) who presented chemical evolution models that can explain the observations with Wolf-Rayet stars and intermittent star formation.

However, GN-z11 is already an evolved galaxy with stellar mass of $\sim 10^9$ $M_\odot$, and to understand the formation and evolution of these galaxies, it is important to simulate them in a cosmological context. This work is the very first attempt to include these chemical enrichment sources self-consistently in cosmological hydrodynamical simulations.

2. Method

We use cosmological hydrodynamical simulations based on the `GADGET-3` code (Springel et al. 2005) in a comoving $25h^{-1}$Mpc cubic box with periodic boundary conditions. The number of gas and dark matter particles are $N_{\rm gas} = N_{\rm DM} = 320^3$, with mass $M_{\rm gas} = 9.34 \times 10^6$ $M_\odot$ and $M_{\rm DM} = 5.1 \times 10^7$ $M_\odot$. We use the same initial conditions and physical processes as in Ibrahim and Kobayashi (2024): metallicity-dependent radiative cooling, star formation, stellar feedback, black hole physics (Kobayashi 2004; Taylor and Kobayashi 2014), and chemical enrichment (Kobayashi et al. 2020). One of the most uncertain phenomena remains supernova feedback.

To constrain supernova feedback in our simulations, we study its impact on the mass–metallicity relation and metallicity radial gradients. In Ibrahim and Kobayashi (2024), we studied the dependence of mass–metallicity relations (MZRs) on supernovae feedback in a $10h^{-1}$Mpc volume up to $z = 3$, for that we implemented and compared four models of supernova feedback to our cosmological simulations: the thermal feedback which consists of pure thermal energy distributed to the neighbour gas particles, the stochastic feedback (similar to Dalla Vecchia and Schaye 2012) stochastically distributing thermal energy to a random number of particles (we use the probability parameter f=50), the kinetic feedback (Navarro and White 1993) partially converting the thermal energy into kinetic energy (the fraction parameter f=1%), and the mechanical feedback (Hopkins et al. 2014) which accounts the physics during the different phases of the supernova expansion (the fraction parameter f=1%).

In Section 3, we constrain supernova feedback in our cosmological simulations using stellar and gas-phase MZRs up to $z = 5$, as well as gas-phase metallicity gradients (Ibrahim and Kobayashi, in prep). Section 4 presents our first results for elemental abundances with our new nucleosynthesis yields including Population III (Pop III) stars.

3. Metallicity evolution

The top panels of Figure 1 show the V-band luminosity-weighted stellar MZRs for the thermal (blue, solid line), stochastic (orange, dash-dotted), kinetic (green, dashed) and mechanical (red, dense dash-dotted) feedback models from $z = 0.7$ to 5. The lines show the medians of all individual galaxies identified in our simulations, while the shaded areas show the 1σ scatter. The dotted lines and symbols are observational data. We use a solar metallicity $Z_\odot = 0.015$.

At $z = 0.7$, our kinetic model seems too strong and does not produce enough metals to explain the observations (see Ibrahim and Kobayashi 2024). Despite no overlap in the mass range, this model seems consistent with data from Gallazzi et al. (2014). However, we note its large error bar of ~ 1 dex. The thermal and stochastic models overproduce metals compared to the mechanical feedback for low-mass galaxies ($M < 10^{10} M_\odot$), which result in significantly higher metallicities than in Gallazzi et al. (2014) for massive ones. These models suggest that the stellar mass–metallicity relation has not significantly evolved between $z \sim 0.7$ and $z \sim 3$, at least at the massive end, contrary to the kinetic model. At $z = 3$, the UV observations from Cullen et al. (2019) for Fe abundances are shifted by +0.5 dex taking account of [O/Fe] and seem to nicely match our kinetic model, but again, we note the large error bar of ~ 1 dex for this model. At $z = 5$, the kinetic model only produces two low-mass galaxies, which explains the absence of MZR for this

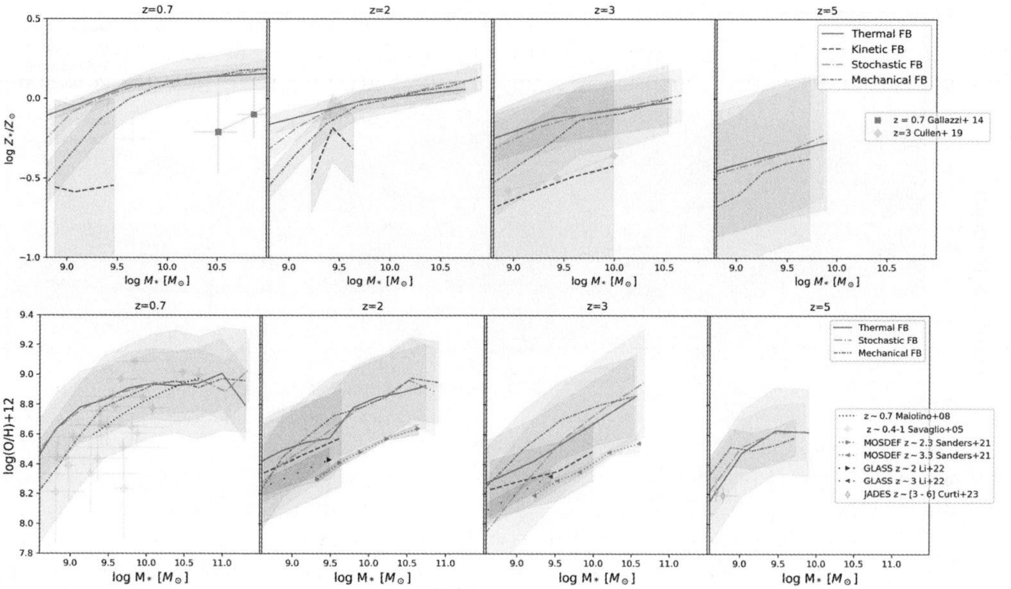

Figure 1. Top panel: Redshift evolution of the luminosity-weighted stellar MZRs of galaxies for the thermal (blue, solid), stochastic (orange,dashed), and mechanical (red, dash-dotted) feedback models. The lines are for the medians, and the shaded areas show the 1σ scatter. Observational data are taken from Gallazzi et al. 2014 ($z = 0.7$), and Cullen et al. 2019 ($z = 3$, with $+0.5$ dex shift for [O/Fe]). Bottom panel is the same but for the SFR-weighted gas-phase MZRs. Observational data (dotted with different symbols) are from Sanders et al. 2021 ($z \sim 2-3$), Li et al. 2022 ($z \sim 2-3$), and Curti et al. 2024 ($z \sim 3-6$).

model. At high redshifts, we retrieve the mass–metallicity relation for thermal, stochastic and mechanical models, with the metallicity at a given mass being lower by ~ 0.4 dex from $z = 5$ to $z = 0.7$. The thermal feedback always produces slightly more metals than the other models for $\mathrm{M} < 10^{10}\mathrm{M}_\odot$, and the kinetic feedback results in significantly lower stellar metallicities than in the other models. Overall, the supernova feedback has a more significant impact on the metallicity at the low-mass end, where low-mass galaxies eject more metals into the intergalactic medium (Kobayashi et al. 2007).

The bottom panel of Figure 1 is the same but for the gas-phase MZRs with SFR-weighted gas oxygen abundance from $z = 0.7$ to 5 for the same four feedback models (see Ibrahim and Kobayashi 2024 for the details). The lines show the medians of individual galaxies in our simulations. The dotted lines and symbols represent observational data at various redshifts, some of which may be suffered by the small sample as already discussed. All observational data have been converted for the Kroupa IMF.

At $z = 0.7$, our kinetic feedback has no star-forming galaxies, which explains the absence of MZR. While our mechanical feedback model most agrees with observed data from Maiolino et al. (2008) and Savaglio et al. (2005). With this resolution, we have a limited sample, but we still retrieve the MZR at all redshifts with thermal, stochastic and mechanical. At higher redshifts, the metallicities at a given mass are considerably lower for the kinetic feedback due to the suppression of star formation, leading to few metal-poor low-mass galaxies. At $z = 5$, the sample of galaxies with the kinetic feedback is too limited to show the MZR. We find a significant evolution from $z \sim 5$ to 0.7 for the gas-phase MZR in the the thermal, stochastic and mechanical feedbacks (~ 0.2 dex). At higher redshifts, the MOSDEF (Sanders et al. 2021) and GLASS survey with NIRISS slitless spectroscopy on JWST (Li et al. 2022) showed ~ 0.1 and ~ 0.07 dex evolution from

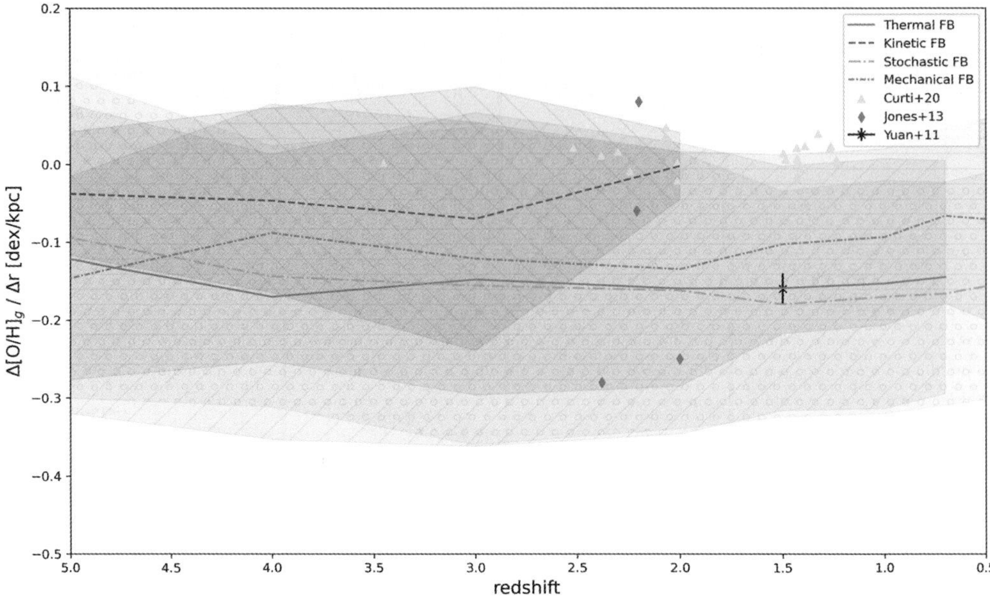

Figure 2. SFR-weighted gas-phase metallicity gradient as a function of redshift, for all galaxies in our simulations with the thermal (blue solid line), kinetic (green dashed), stochastic (orange dash-dotted), and mechanical (red dense dash-dotted) feedback models. The shaded areas are 1σ scatter. The symbols are observational data from Yuan et al. 2011 (black cross) using AO-assisted spectroscopy OSIRIS on Keck II on a face-on spiral galaxy at $z \sim 1.5$, Jones et al. 2013 (magenta diamonds) using AO-assisted spectroscopy OSIRIS on Keck on gravitationally lensed systems, and Curti et al. 2020 (cyan triangle) using KMOS KLEVER survey on 42 gravitationally lensed galaxies.

$z \sim 2$ to $z \sim 3$, respectively, which is larger than in all our models (~ 0.07 dex), except for the kinetic. The kinetic feedback model fits well with the laters, but as discussed previously, this model is underproducing stars, so this matching does not necessarily support the kinetic feedback of supernovae. At $z = 5$, the JADES survey (Curti et al. 2024) with JWST/NIRSpec at $3 < z < 6$ for low mass galaxies show a nice match with our thermal model.

Figure 2 shows the SFR-weighted gas-phase metallicity gradients as a function of redshift for all our galaxies with the thermal (blue solid line), kinetic (green, dashed), stochastic (orange, dash-dotted), and mechanical (red, dense dash-dotted) feedback models. The gradient is measured within $2R_e$ where R_e is the effective radius of each galaxy. We show the redshift evolution from $z = 5$ up to $z = 0.5$ for the stochastic and mechanical models, to $z = 0.7$ for the thermal model, and to $z = 2$ for the kinetic feedback which stops forming stars after $z = 2$ with this resolution. The shaded areas show the 1σ scatter.

The symbols represent observational data from Curti et al. 2020 (cyan triangle), Yuan et al. 2011 (black cross), Jones et al. 2013 (magenta diamonds). These observational data nicely match our gradients with the kinetic feedback near $z \sim 2$. The gradients with the mechanical feedback is flatter than those with thermal and stochastic feedback, but with all three supernovae feedback models, the gradients remain mostly negative at all redshifts, suggesting that galactic centers are more metal-rich than the outskirts. A detailed analysis of both of the stellar and gas-phase metallicity gradients will be discussed in Ibrahim and Kobayashi, in prep.

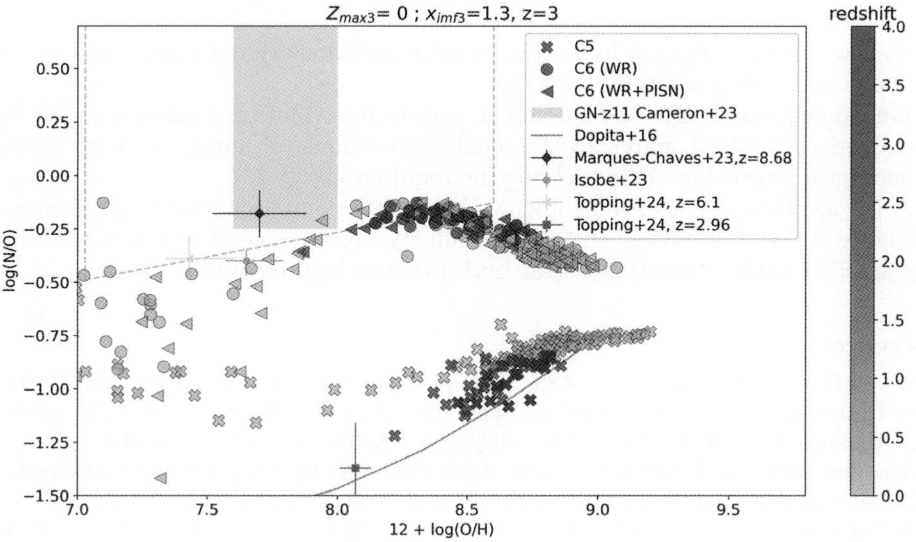

Figure 3. The evolution of SFR-weighted gas-phase N/O abundance ratios compared to oxygen abundances with Pop III nucleosynthesis yields from WR (circles), WR and PISN (triangles) and without (cross) (including faint supernovae)

4. Elemental abundance

To explain the N/O ratios of very high redshift galaxies, we use the latest nucleosynthesis yields from Kobayashi et al. 2020, but including PISN (Takahashi et al. 2018), WR stars (Limongi and Chieffi 2018), and faint supernovae for Pop III stars, as in Kobayashi and Ferrara (2024). We have varied many parameters for the Pop III contributions in our simulations but show the abundance ratios with three different cases in Figure 3, all with a constant Kroupa IMF slope (Kroupa 2008) between the lower and upper mass limits ($[m_\ell, m_\mathrm{u}]$). For the comparison to our previous “C5” simulations in Section 3, we use thermal feedback only.

In Figure 3 we show our simulation C5 without nucleosynthesis yields from PISN or WR stars (cross symbols) from $z = 0$ (dark gray), to $z = 4$ (light gray). This gives a nice match with local observational data (solid line) from Dopita et al. (2016). This observed local relation has been explained as follows. (i) The plateau at low metallicities is caused by rotating massive stars in one-zone galactic chemical evolution models (Chiappini et al. 2006), *or* inhomogeneous enrichment from asymptotic giant branch (AGB) stars in cosmological simulations as explained in Vincenzo and Kobayashi (2018). (ii) The increasing trend toward higher metallicities is caused by the metallicity dependence of N yields in all mass range of stars *and* the delayed enrichment from AGB stars.

We also show our model C6 including Pop III star nucleosynthesis yields from rotating WR stars (circles) with $[0.01{,}120]\mathrm{M}_\odot$, and another model C6 including both rotating WR stars and PISNe (triangles) with $[15,\ 300]\mathrm{M}_\odot$. These C6 simualtions produce a significantly different N/O ratios, mainly caused by the rotating star models. In massive stars, rotation enhances mass loss through stellar winds, ejecting elements like nitrogen and carbon and increasing the number of WR stars (e.g. Limongi and Chieffi 2018). As a result, the presence of rotating WR stars (at all metallicities) enhances N/O abundance ratios in galaxies (at all redshifts). The stellar rotation highly impacts the chemical enrichment history of galaxies, much more than the presence of PISNe.

4.1. *Conclusion*

We show our cosmological hydrodynamical simulations that include the latest nucleosynthesis yields. Our main findings are:

• Our analysis of MZRs and metallicity gradients evolution suggest that mechanical feedback gives the best match to the local observations, although stronger supernovae feedback such as our kinetic model may be required at $z > 2$.

• Nucleosynthesis yields from rotating WR stars in our simulations can enhance N/O abundance ratios in galaxies, and the metallicity dependence of stellar rotations might be required to explain observations at both low and high redshifts.

References

Andrew J. Bunker et al. *A&A*, 677:A88, September 2023. doi: 10.1051/0004-6361/202346159
Marco Castellano et al. *ApJ*, 972(2):143, September 2024. doi: 10.3847/1538-4357/ad5f88
C. Charbonnel et al. *A&A*, 673:L7, May 2023. doi: 10.1051/0004-6361/202346410
C. Chiappini et al. *A&A*, 449(2):L27–L30, April 2006. doi: 10.1051/0004-6361:20064866
F. Cullen et al. *MNRAS*, 487(2):2038–2060, August 2019. doi: 10.1093/mnras/stz1402
Mirko Curti et al. *MNRAS*, 492(1):821–842, February 2020. doi: 10.1093/mnras/stz3379
Mirko Curti et al. *A&A*, 684:A75, April 2024. doi: 10.1051/0004-6361/202346698
Claudio Dalla Vecchia and Joop Schaye. *MNRAS*, 426(1):140–158, October 2012
Michael A. Dopita et al. *Ap&SS*, 361:61, February 2016. doi: 10.1007/s10509-016-2657-8
Anna Gallazzi et al. *ApJ*, 788(1):72, June 2014. doi: 10.1088/0004-637X/788/1/72
Philip F. Hopkins et al. *MNRAS*, 445(1):581–603, November 2014. doi: 10.1093/mnras/stu1738
Dyna Ibrahim and C. Kobayashi. *MNRAS*, 527(2):3276–3290, January 2024. doi: 10.1093/mnras/stad3313
Yuki Isobe et al. *ApJ*, 959(2):100, December 2023. doi: 10.3847/1538-4357/ad09be
Tucker Jones et al. *ApJ*, 765(1):48, March 2013. doi: 10.1088/0004-637X/765/1/48
Chiaki Kobayashi. *MNRAS*, 347(3):740–758, January 2004. doi: 10.1111/j.1365-2966.2004.07258.x
Chiaki Kobayashi and Andrea Ferrara. *ApJ*, 962(1):L6, February 2024. doi: 10.3847/2041-8213/ad1de1
Chiaki Kobayashi, Volker Springel, and Simon D. M. White. 376(4):1465–1479, mar 2007
Chiaki Kobayashi, Amanda I. Karakas, and Maria Lugaro. *ApJ*, 900(2):179, September 2020
Pavel Kroupa. In Sverre J. Aarseth, Christopher A. Tout, and Rosemary A. Mardling, editors, volume 760, page 181, 2008. doi: 10.1007/978-1-4020-8431-7_8
Mingyu Li et al., 2022
Marco Limongi and Alessandro Chieffi. *ApJS*, 237(1):13, July 2018. doi: 10.3847/1538-4365/aacb24
R. Maiolino et al. *A&A*, 488(2):463–479, September 2008. doi: 10.1051/0004-6361:200809678
R. Marques-Chaves et al. *A&A*, 681:A30, January 2024. doi: 10.1051/0004-6361/202347411
Chris Nagele and Hideyuki Umeda. *ApJ*, 949(1):L16, May 2023. doi: 10.3847/2041-8213/acd550
Devesh Nandal et al. *A&A*, 683:A156, March 2024. doi: 10.1051/0004-6361/202348035
J. F. Navarro and S. D. M. White. *MNRAS*, 265:271, November 1993. doi: 10.1093/mnras/265.2.271
Ryan L. Sanders et al. *ApJ*, 914(1):19, June 2021. doi: 10.3847/1538-4357/abf4c1
S. Savaglio et al. *ApJ*, 635(1):260–279, December 2005. doi: 10.1086/497331
D. Schaerer et al. *A&A*, 687:L11, July 2024. doi: 10.1051/0004-6361/202450721
Peter Senchyna et al. *ApJ*, 966(1):92, May 2024. doi: 10.3847/1538-4357/ad235e
Volker Springel et al. *MNRAS*, 361(3):776–794, August 2005. doi: 10.1111/j.1365-2966.2005.09238.x
Koh Takahashi et al. *ApJ*, 857(2):111, April 2018. doi: 10.3847/1538-4357/aab95f
P. Taylor and C. Kobayashi. *MNRAS*, 442(3):2751–2767, August 2014. doi: 10.1093/mnras/stu983
Michael W. Topping et al. *MNRAS*, 529(4):3301–3322, April 2024a. doi: 10.1093/mnras/stae682

Michael W. Topping et al. *arXiv e-prints*, art. arXiv:2407.19009, July 2024b. doi: 10.48550/arXiv.2407.19009

F. Vincenzo and C. Kobayashi. *MNRAS*, 478(1):155–166, July 2018. doi: 10.1093/mnras/sty1047

T. T. Yuan, L. J. Kewley, A. M. Swinbank, J. Richard, and R. C. Livermore. *ApJ*, 732(1):L14, May 2011. doi: 10.1088/2041-8205/732/1/L14

The First Chapters of Our Cosmic History with JWST
Proceedings IAU Symposium No. 391, 2025
T. Nanayakkara & M. Maseda, eds.
doi:10.1017/S1743921324002795

JWST observations of ALMA [O III] 88 μm emitters in the epoch of reionization

Takuya Hashimoto

Division of Physics, Faculty of Pure and Applied Sciences, University of Tsukuba, Tsukuba, Ibaraki 305-8571, Japan

Tomonaga Center for the History of the Universe (TCHoU), Faculty of Pure and Applied Sciences, University of Tsukuba, Tsukuba, Ibaraki 305-8571, Japan
email: `hashimoto.takuya.ga@u.tsukuba.ac.jp`

Abstract. Recent ALMA studies have detected rest-frame far-infrared emission lines in star-forming galaxies at redshift (z) $\sim 6 - 9$. To investigate detailed properties of high-z ALMA [O III] 88 μm emitters, we performed JWST observations with NIRCam and NIRSpec IFU (GO-1840). In the NIRCam observations, six broad- and/or medium-band filters were used to trace rest-frame UV to optical stellar continuum emission, while NIRSpec IFU observations provide information on rest-frame optical emission lines. The powerful combination of JWST and ALMA in this project has revealed a matured stellar population in the core region of the most distant protocluster, A274z7p9OD, at $z = 7.9$. We also present our results of gas-phase metallicity from a direct-temperature method based on a joint analysis of the optical and far-infrared [O III] lines, which will form the basis for similar studies combining JWST and ALMA data.

Keywords. galaxies: high-redshift, galaxies: ISM, galaxies: star formation.

1. Introduction

Understanding properties of galaxies at redshift of $z \gtrsim 6$ is crucial to study galaxy formation and evolution as well as the process of cosmic reionization. With the advent of the James Webb Space Telescope *(JWST)*, it has become possible not only identifying distant galaxies well before the completion of reionization, but also studying their detailed properties such as their stellar populations and chemical enrichment by probing rest-frame ultraviolet to optical emission (e.g., Curtis-Lake et al. 2023). ALMA has a unique capability to study rest-frame far-infrared properties of high-z galaxies such as the far-infrared fine structure lines of [C II] 158 μm and [O III] 88 μm, and dust thermal continuum emission.

2. Overview of the *JWST* GO-1840 program

As part of the Reionization and the ISM/Stellar Origins with JWST and ALMA (RIOJA) project, we are conducting follow-up observations of twelve [O III] 88 μm emission line galaxies at $z = 6.0 - 8.3$ identified by ALMA (e.g., Hashimoto et al. 2018) with NIRCam and NIRSpec IFU onboard JWST (GO-1840, PIs: J. Álvarez-Márquez and T. Hashimoto, see also Sugahara et al. 2024). The sample selection was based on a complete sample of fourteen ALMA [O III] 88 μm emitters at $z = 6.0 - 9.1$, two of which (MACS1149-JD1 and MACS0416-Y1) were removed because they were included in several GTO programs. The targeted galaxies have a high dynamic range in the ultraviolet absolute magnitudes ranging from −23 to −20, which covers the typical to bright-end Lyman break galaxies.

The NIRCam observations use six filters carefully selected for the targeted galaxies on the individual basis. The shorter three wave bands probe the rest-frame ultraviolet continuum that traces the young massive stars. The two longer wave bands probe the wavelengths around the Balmer break, and the longest wave band probes the rest-frame optical continuum contaminated by strong optical emission lines such as [O III] 5008 Å and Hα. The NIRSpec IFU observations adopt a grating/filter pair of G395H/F290LP (F395MF290LP) that produces a cube with a spectral resolution of $R \sim 2700$ ($R \sim 1000$) in a wavelength range of 2.87 − 5.27 μm for the galaxies at $z \sim 6.0 - 6.8$ and 7.9 ($z \sim 7.1$). The NIRSpec IFU data provide strong optical emission lines such as [O II] 3728, 3739 Å, Hβ, [O III] 4960, 5008 Å, and Hα.

Combination of the NIRCam and NIRSpec IFU data is crucial to accurately measure the optical line fluxes that contaminate the NIRCam images, thereby performing accurate SED fitting of the targeted galaxies. With these data, we plan to examine the stellar buildup of the targeted galaxies, how the older stellar populations are prevalent among the high-z [O III] 88 μm emitters. Because all the targeted galaxies have information on dust from ALMA regardless it is detected or not, we can examine spatial distributions of stars, ionized gas, and dust in these galaxies. With these spatial information, we plan to examine whether these early galaxies have formed as compact clumps or in a large extended gaseous disks. These two will address our first question: the prevalence of older stellar populations in the early galaxies. With the NIRSpec IFU data, we plan to obtain important ISM parameters such as the ionization parameter, gas-phase metallicity, electron density, and electron temperature. In this respect, the combination of ALMA [O III] 88 μm and *JWST* optical [O III] 4960, 5008 Å emission lines has potential to constrain the electron temperature even without the fainter [O III] auroral line at 4364 Å (Nakazato et al. 2023), which is crucial to estimate the gas-phase metallicity accurately. These ISM parameters will allow us to understand the second question: physical origins of the high ALMA [O III] 88 μm-to-[C II] 158 μm luminosity ratios, how the galaxy population has contributed to the reionization process.

3. Results and Discussion

3.1. *1. Matured stellar population in a protocluster at $z \sim 7.9$, A2744-z7p9OD*

An interesting candidate protocluster, A2744-z7p9O, at $z \approx 7 - 8$ behind the strong lensing cluster Abell 2744 was reported by independent groups (e.g., Laporte et al. 2014) based on photometric redshift deduced from *HST* data taken as a part of Hubble Frontier Fields Survey (PI: Lotz) in conjunction with *Spitzer*/IRAC data. Remarkably, the proto-cluster candidate includes as many as eleven (eleven) member galaxy candidates at $7 < z < 8$ ($8 < z < 9$) within just a single field of view of *HST* WFC3. Among these, particularly interesting is the presence of "quintet" galaxy candidates at $z \sim 8$ named A2744-YD1, YD4, YD6, YD7, and zD1 in an apparent distance of just ∼ 3" × 3". Recently, Morishita et al. (2023) have reported the spectroscopic redshift of A2744-z7p9O by identifying seven galaxies at $z = 7.88$ with [O III] 5007 Å. The observations were made by JWST ERS and DDT programs using the micro shutter array (MSA) mode with a prism ($R \sim 100$) and/or medium-resolution ($R \sim 1000$) configurations.

In our GO1 program, we targeted the "quintent" galaxy candidates with NIRSpec IFU with a high spectral resolution mode (Hashimoto et al. 2023). Among the five member galaxy candidates, A2744-YD1, YD4, YD6, YD7, zD1, we clearly detected three member galaxies, A2744-YD1, YD4, YD7, and one additional serendipitous galaxy, A2744-S1, at $z \sim 7.88$ within a single field-of-view of IFU (left panel of Figure 1). A2744-S1 was not

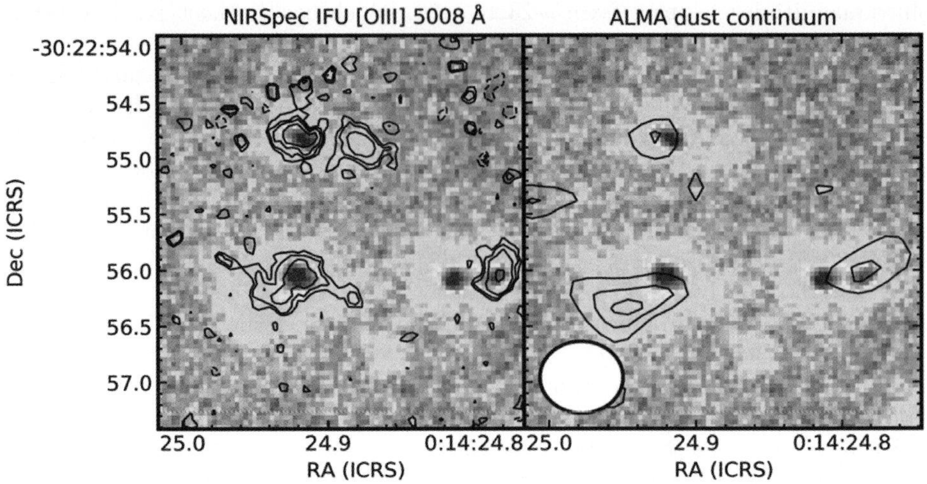

Figure 1. The colored background image shows NIRCam F444W around the "quintet" region. (Left) Black contours show the [O III] 5008 Å line map. (Right) Black contours show the ALMA Band 6 dust continuum map at the rest-frame wavelength of ~ 133 μm. The figure is quoted from Hashimoto et al. (2023).

recognized as a $z \sim 8$ candidate in the previous literatures based on the HST and Spitzer data. This clearly demonstrates the power of IFU observations to understand the true structure of a proto-cluster in a less biased manner.

With the ALMA Band 6 archival data, we have also detected dust continuum emission from three out of the four member galaxies, A2744-YD1, YD4, YD7 (right panel of Figure 1). This is the first detection of dust continuum emission from member galaxies of a protocluster at such a high redshift. The presence of dust in the member galaxies are consistent with their red UV continuum slopes (β) obtained from NIRCam imaging data. At a given UV absolute magnitude, these three galaxies clearly show higher β compared to typical galaxies at similar redshifts. This demonstrates an accelerated star formation activity in the core region of a protocluster.

3.2. *2. Spatially-resolved and integrated properties of a $z \sim 6.8$ galaxy, COS-2987*

COS-2987 is a bright Lyman break galaxy originally identified by ALMA's spectroscopy in [C II] 158 μm and [O III] 88 μm (Witstok et al. 2022). In our GO1 program, we targeted the galaxy with both NIRCam and NIRSpec IFU in the high spectral resolution mode (K. Mawatari et al. in preparation; M. Usui et al. in preparation). We successfully detected key strong emission lines such as [O II], Hβ, [O III] 4364 Å, [O III] 4960, 5008 Å, and Hα. With these data, the spatially-integrated spectrum suggests that COS-2987 is a dust-poor, low-metallicity ($\sim 10\%$ solar) galaxy. However, the spatially-resolved spectrum shows that the galaxy has a non-negligible amount of dust in a region where intense star formation activity is ongoing. We have also combined ALMA's [O III] 88 μm data in addition to the aforementioned optical emission lines. Remarkably, we have found that the [O III] 88 μm flux is too high at a given parameter set of electron density and temperature derived from rest-frame optical emission lines. This may be due to the fact that the far-infrared [O III] 88 μm has a critical density against electron, $n_{\rm crit.} \sim 500$ cm^{-3}, which is much lower than those of the optical [O III] lines. The far-infrared [O III] 88 μm line may trace diffuse ionized gas that cannot be easily traced by strong optical emission lines alone.

During the presentation, there was an interesting question regarding the presence of AGN activity in the sample. This can be tested by using strong rest-frame optical emission lines (e.g., BPT diagram), but this does not show any strong evidence of AGN activity. The galaxies also do not show a broad component in the Balmer lines. Furthermore, the galaxies are not detected in X-ray or radio catalogues. Altogether, these indicate that the galaxies do not show any strong evidence of AGN activity.

References

Curtis-Lake, E., Carniani, S., Cameron, A., et al. 2023, *NatAs*, 7, 622, doi: 10.1038/s41550-023-01918-w

Sugahara, Y., Álvarez-Márquez, J., Hashimoto, T., et al. 2024, arXiv:2403.17133. doi:10.48550/arXiv.2403.17133

Hashimoto, T., Laporte, N., Mawatari, K., et al. 2018, *Nature*, 557, 392, doi: 10.1038/s41586-018-0117-z

Nakazato, Y., Yoshida, N., & Ceverino, D. 2023, *ApJ*, 953, 140, doi: 10.3847/1538-4357/ace25a

Laporte, N., Streblyanska, A., Clement, B., et al. 2014, *A&A*, 562, L8, doi: 10.1051/0004-6361/201323179

Morishita, T., Roberts-Borsani, G., Treu, T., et al. 2023, *ApJL*, 947, L24, doi: 10.3847/2041-8213/acb99e

Hashimoto, T., Álvarez-Márquez, J., Fudamoto, Y., et al. 2023, *ApJL*, 955, L2, doi: 10.3847/2041-8213/acf57c

Witstok, J., Smit, R., Maiolino, R., et al. 2022, *MNRAS*, 515, 1751, doi: 10.1093/mnras/stac1905

The First Chapters of Our Cosmic History with JWST
Proceedings IAU Symposium No. 391, 2025
T. Nanayakkara & M. Maseda, eds.
doi:10.1017/S1743921324002771

Tracing Outflows from Stellar Feedback in the Early Universe with Lyman-α

James Nianias[1], Jeremy Lim[1], Yik Lok Wong[2], Gordon Wong[1], Ishika Kaur[1] and Wenjun Chen[1]

[1]Department of Physics, University of Hong Kong, Pokfulam Road, Hong Kong
email: nianias@hku.hk

[2]Department of Physics and Astronomy, The University of Manchester, Oxford Road, Manchester M13 9PL, UK

Abstract. Blind spectroscopy of massive lensing galaxy clusters with MUSE has revealed large numbers of gravitationally-lensed Lyman-α emitters exhibiting asymmetric profiles at $2.9 \leq z \leq 6.7$, suggesting abundant outflows from low-mass star-forming galaxies in the early universe. Are these primaeval galaxies experiencing their first bursts of star formation, or established galaxies experiencing rejuvenation? With JWST rest-frame optical/NIR continuum imaging now available for many of these objects, we can search for older stellar populations. Here, we search for spectroscopic confirmation of outflows from these galaxies, finding a few high-signal-to-noise cases in which blueshifted interstellar absorption lines are detected. Next, we analyse the star formation histories with combined HST + JWST photometry. We find most them to be well characterised by very young, low metallicity stellar populations. However, despite the rest-frame optical/NIR coverage of JWST, we cannot place strict upper bounds on the mass in old stars (age > 100 Myr).

Keywords. Early universe, galactic outflows, star formation, Lyman-α emitters

1. Introduction

Since their first detections in the mid-1990s, Lyman-α emitters (LAEs) have become important tools with which to study the early universe. With specific star formation rates (sSFRs) on or above the main sequence and low dust content (Iani et al. 2024), much of the radiation from beyond the Lyman limit ($\lambda < 912$ Å) is reprocessed into the Lyman-α line, which consequently can have extremely large equivalent width (EW). This makes LAEs easy to detect in narrowband imaging (e.g. Ouchi et al. 2003) as well as blind spectroscopy (e.g. Bacon et al. 2017) despite their generally low masses ($10^6 \lesssim M_\star/\mathrm{M}_\odot \lesssim 10^9$; Iani et al. 2024). Spectroscopic detection is also made easier by the fact that Lyman-α's resonant nature couples it to the kinematics of the interstellar medium (ISM), producing a distinctive and easily identifiable asymmetric line profile. Line profiles of this kind can be reproduced using expanding shells of gas with varying outflow velocities, neutral covering fractions, and dust content (Ahn 2004; Verhamme et al. 2006; Orsi et al. 2012). Stacked spectra of LAEs and Lyman break galaxies (LBGs) at high redshift, as well as some individual gravitationally-lensed galaxies, also exhibit blueshifted interstellar absorption features, cementing the link between asymmetric Lyman-α profiles and outflows (Franx et al. 1997; Shapley et al. 2003; Berry et al. 2012; Jones et al. 2012).

Richard et al. (2021) published spectro-photometric catalogues based on blind spectroscopy of massive lensing clusters with the Multi-Unit Spectroscopic Explorer (MUSE) and imaging with the Hubble Space Telescope (HST). These catalogues cover a total of

12 lensing clusters, and include ~ 1000 lensed images of LAEs with redshifts from 2.9 to 6.7 (i.e. into the epoch of reionization). The magnification provided by gravitational lensing (up to a factor of ~ 100) enables detections of intrinsically fainter LAEs than would otherwise be possible, and also permits measurements of the physical sizes of the UV continuum counterparts with HST down to the scale of just a few tens of pc (Claeyssens et al. 2022) – the size of individual globular clusters (Vanzella et al. 2017). The nature of these faint and compact systems is still uncertain: are we seeing star forming regions in established galaxies, or primaeval star bursts in low-mass dark matter halos? With the enhanced wavelength coverage, sensitivity, and spatial resolution afforded by the James Webb Space Telescope (JWST), we can begin to search for signs of older stellar populations in these objects by observing the rest-frame optical/near-IR (NIR).

In Section 2, we use the MUSE spectra to confirm outflow signatures in these objects, comparing the Lyman-α, optically thin high-ionization nebular emission, and interstellar absorption lines. To answer the question of whether these objects host older stellar populations, we perform SED fitting of the HST + JWST photometry, presented in Section 3.

2. MUSE Spectroscopy

We extracted 1D spectra for each source from the corresponding MUSE cubes. We chose only those sources with Lyman-α detected with a signal-to-noise ratio (SNR) of 3 or better according to the catalogues of Richard et al. (2021). We then fitted analytic line profiles to each spectral line in listed in the catalogues: Lyman-α, nebular emission, and interstellar absorption lines. For the nebular emission and interstellar absorption lines, we used gaussian profiles. However, in order to model the asymmetric shape of the Lyman-α line, we instead used an asymmetric gaussian profile as in (Shibuya et al. 2014). Since Lyman-α lines sometimes also have weaker secondary peaks blueshifted relative to the primary, we attempted to fit a blueshifted component as well. In cases where the blueshifted peak (henceforth blue peak) was statistically significant and well-resolved from the primary (henceforth red peak), we considered it a detection.

Having fitted all lines in each spectrum, We searched for spectra with kinematic signatures of outflows. First, we selected those with statistically significant nebular emission lines and absorption lines. The nebular emission lines are valuable as they allow us to measure the systemic redshift of the star-forming regions responsible for the Lyman-α emission, which cannot be done using Lyman-α itself due to the aforementioned asymmetric line profiles. The interstellar absorption lines, on the other hand, can reveal outflowing gas screening the continuum source if they are blueshifted. Confident identification of genuine emission and absorption features (besides Lyman-α) is complicated by the presence of residual atmospheric features in the MUSE spectrum. However, we identified four confident examples of Lyman-α emitters with both emission and absorption lines. Comparing these lines in velocity space with the Lyman-α profile, we find clear evidence of outflows: the absorption is blueshifted relative to the systemic redshift, or at least extends to more blueshifted velocities than the nebular emission, while the Lyman-α is redshifted, with a skew towards higher velocities. We show a comparison of the Lyman-α, emission, and absorption lines from two of these spectra in Figure 1. These kinematics agree with those found in studies of other LAEs/LBGs (e.g. Shapley et al. 2003; Jones et al. 2012; Berry et al. 2012).

All the spectra we find with unmistakable absorption features also have singly-peaked Lyman-α emission (i.e. no blue peaks). While the number of spectra is small, some of the Lyman-α profiles have extremely high SNR (e.g. panels (a) and (b) of Figure 1), suggesting that this is not merely a consqeuence of blue peaks being below the noise level. Qualitatively, this fits with the picture presented in Erb (2015), in which there is

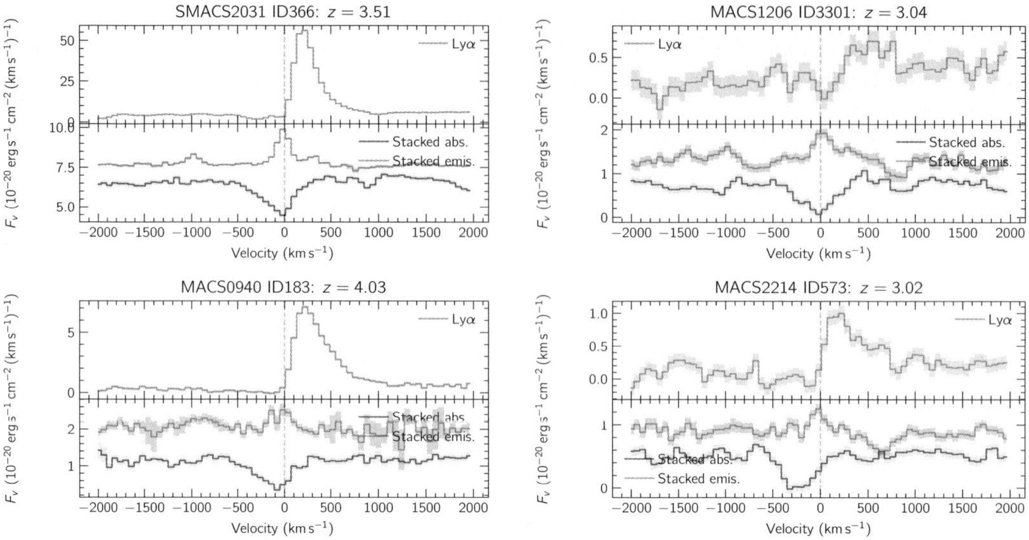

Figure 1. Examples of MUSE spectra from Lyman-α emitters where we find strong evidence for low-ionization interstellar absorption features. The Lyman-α profiles are shown in the upper panels in blue, and the stacked emission and absorption lines in the lower panels in red and black respectively. The spectra show clear evidence for outflows in that the absorption lines appear either blueshiftee with respect to the emission, or else extend to more blueshifted velocities, while the Lyman-α lines are asymmetric and redshifted relative to the systemic.

a greater covering fraction of neutral gas in single-peaked Lyman-α spectra, which also gives rise to the absorption lines (which are mostly low-ionization species associated with neutral gas).

3. HST+JWST SED fitting

Having established the presence of outflows in at least some of these sources, we turn to the question of their star formation histories. Studies of selected individual objects included in this sample with HST imaging suggest extremely young ages, low metallicities, compact sizes, and high specific star formation rates (Vanzella et al. 2017). However, with the HST filters providing only coverage of the rest-frame UV (or, at best, optical for the lowest-redshift sources), it has been impossible to place strong constraints on the mass in any older stellar population. With JWST/NIRCam imaging of two of the clusters included in the Richard et al. (2021) sample now available, providing up to rest-frame NIR coverage for sources with $z \lesssim 4.0$, we can now search for signatures of older stellar populations in these objects.

We obtained JWST/NIRCam images of two clusters: MACS J0416.1-2403 (via the PEARLS collaboration; Windhorst et al. 2023) and Abell 2744 (from Data Release 3 of the UNCOVER survey; Suess et al. 2024), which we refer to as MACS0416 and A2744, respectively, for the remainder of this paper. The PEARLS images of MACS0416 includes mostly broad-band and one medium-band filter (F090W, F115W, F200W, F277W, F356W, F410M, F444W), while the UNCOVER images of A2744 include these plus a number of additional medium-band filters (F140M, F162M, F182M, F210M, F250M, F300M, F335M, F360M, F430M, F460M, F480M) and one additional broad-band filter (F070W). Both of these data sets are complemented by archival broad-band HST ACS and WFC images (F435w, F606W, F814W, F105W, F125W, F140w, F160W). For the images of MACS0416, we aligned the images and matched the PSFs to the image with the coarsest angular resolution, which we found to be the HST/WFC F125W filter.

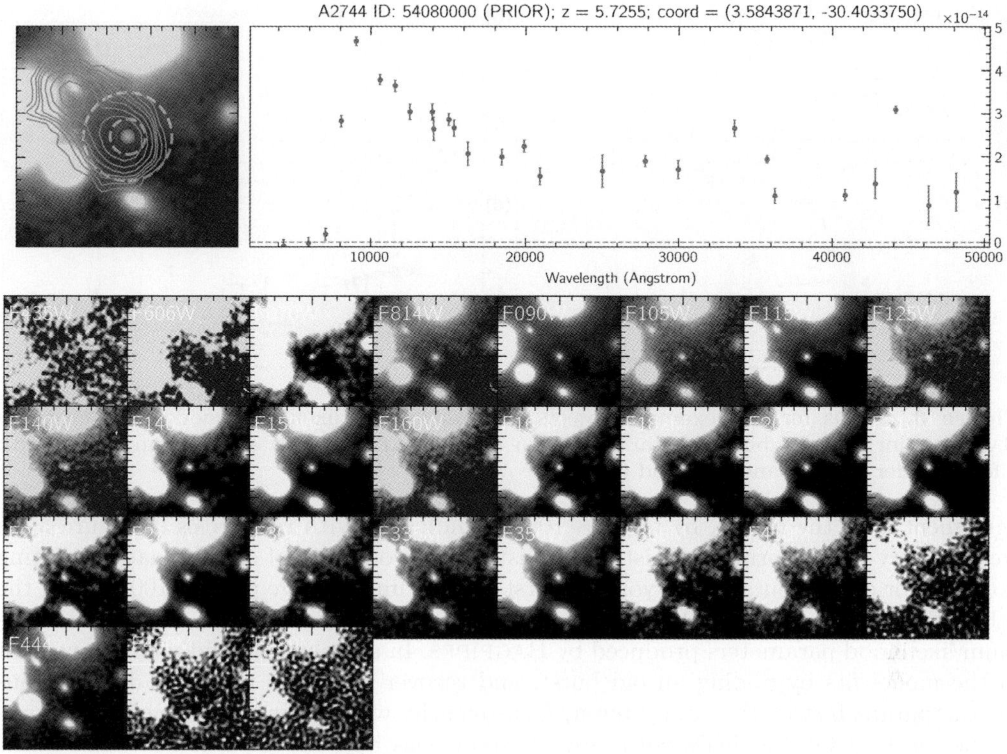

Figure 2. Example object from A2744. Above: stacked image (left) with MUSE Lyman-α contours (orange) and showing aperture with background annulus, and extracted SED (left) with HST measurements in blue and JWST in red. Below: individual postage stamp cutouts in each HST and JWST filter, arranges by wavelength. The morphology appears consistent across all filters.

We then fitted light profiles to the bright, spectroscopically confirmed cluster members (using single Sersics, double Sersics, or core-Sersics), and subtracted them to mitigate contamination of the LAEs' photometry. For the images of A2744, we simply obtained the aligned, PSF-matched, cluster-member-subtracted images directly from the UNCOVER data release†.

As an initial check for older stellar populations, we visually compared the morphology of the sources in the HST and JWST images. In many cases, we find no difference in morphology between the rest-frame UV and optical/NIR, as shown in the example in Figure 2. This demonstrates that, far from being star forming regions embedded in more massive galaxies, many of these objects are as compact as prior studies with HST suggest (e.g. Claeyssens et al. 2022).

We performed apeture photometry on all the most highly magnified images of LAEs ($\mu \geq 10$), using magnifications and positions from the catalogues of Richard et al. (2021). In both clusters, we set the aperture diameter to be 2× the full-width at half-maximum (FWHM) of the respective PSFs. Having extracted SEDs for our objects, we then performed SED modelling using Bagpipes (Carnall et al. 2018). We started by fitting the simplest possible star formation history: a single burst of star formation. We find that in many, if not most, of the spectra, the maximum-likelihood single-burst model is already in good agreement with the observed SEDs as shown in Figure 3(a). On the other hand, we find that some of the UNCOVER SEDs contain some points that are in very poor

† `https://jwst-uncover.github.io/DR3.html`

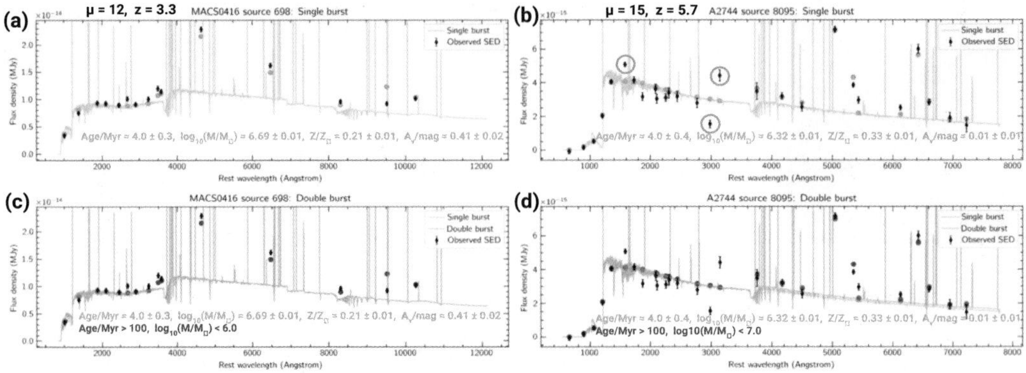

Figure 3. SEDs for two example objects with maximum likelihood models produced by BAGPIPES for a single burst of star formation (panels (a) and (b)), and a young burst with a forced older burst (panels (c) and (d)).

agreement with the model, by up to $\sim 10\sigma$, as indicated with red circles in Figure 3(b). Next, we try a star formation history consisting of two bursts of star formation, setting no age prior on one burst (the "young" burst), and a uniform age prior > 100 Myr on the other (the "old" burst). We show example two-burst models in Figure 3 along with maximum likelihood parameters produced by BAGPIPES. In all cases, we find no improvement in the model fits by adding an old burst, and recover the same posterior distributions for the parameters of the young burst. Consquently, we still find a few points in some of the the UNCOVER SEDs with large discrepencies between data and models. These discrepencies appear largely random, and are not resolved by trying more complex star formation histories (exponential, constant, or power law).

Our SED models show no evidence of older stellar populations in these objects. However, inspecting the mass posteriors for the old bursts, we find a more complex story: while for some objects the mass of the old burst can be constrained to $\sim$ an order of magnitude less than the young burst, as in Figure 3(c), in other cases the upper bounds are much less strict, and can even be higher than the mass in young stars as shown in Figure 3(d). This demonstrates an important point: outshining by the young stellar population can be severe enough in these objects that even JWST/NIRCam observations are relatively insensitive to older stellar populations. To convincingly rule out older generations of stars in these systems, either longer wavelength coverage with JWST/MIRI (e.g. Goovaerts et al. 2023), or else deeper observations with 30 m-class telescopes are required.

4. Acknowledgements

J.L. and J.N. acknowledge support from the Research Grants Council of Hong Kong for conducting this work under the General Research Fund 17312122. We also thank Rogier Windhorst and the rest of the PEARLS collaboration for providing us with early access to the JWST images of MACS0416.

References

Ahn, S.-H. 2004, Singly Peaked Asymmetric Lyα from Starburst Galaxies. *ApJ*, 601(1), L25–L28

Bacon, R., Conseil, S., Mary, D., Brinchmann, J., Shepherd, M., Akhlaghi, M., Weilbacher, P. M., Piqueras, L., Wisotzki, L., Lagattuta, D., Epinat, B., Guerou, A., Inami, H., Cantalupo, S., Courbot, J. B., Contini, T., Richard, J., Maseda, M., Bouwens, R., Bouché, N., Kollatschny, W., Schaye, J., Marino, R. A., Pello, R., Herenz, C., Guiderdoni, B., & Carollo, M. 2017, The MUSE Hubble Ultra Deep Field Survey. I. Survey description, data reduction, and source detection. *A&A*, 608, A1

Berry, M., Gawiser, E., Guaita, L., Padilla, N., Treister, E., Blanc, G. A., Ciardullo, R., Francke, H., & Gronwall, C. 2012, Stacked Rest-frame Ultraviolet Spectra of Lyα-emitting and Continuum-selected Galaxies at 2 ¡ z ¡ 3.5. *ApJ*, 749(1), 4

Carnall, A. C., McLure, R. J., Dunlop, J. S., & Davé, R. 2018, Inferring the star formation histories of massive quiescent galaxies with BAGPIPES: evidence for multiple quenching mechanisms. *MNRAS*, 480(4), 4379–4401

Claeyssens, A., Richard, J., Blaizot, J., Garel, T., Kusakabe, H., Bacon, R., Bauer, F. E., Guaita, L., Jeanneau, A., Lagattuta, D., Leclercq, F., Maseda, M., Matthee, J., Nanayakkara, T., Pello, R., Thai, T. T., Tuan-Anh, P., Verhamme, A., Vitte, E., & Wisotzki, L. 2022, The Lensed Lyman-Alpha MUSE Arcs Sample (LLAMAS). I. Characterisation of extended Lyman-alpha halos and spatial offsets. *A&A*, 666, A78

Erb, D. K. 2015, Feedback in low-mass galaxies in the early Universe. *Nature*, 523(7559), 169–176

Franx, M., Illingworth, G. D., Kelson, D. D., van Dokkum, P. G., & Tran, K.-V. 1997, A Pair of Lensed Galaxies at z = 4.92 in the Field of CL 1358+62. *ApJ*, 486(2), L75–L78

Goovaerts, I., Pello, R., Thai, T. T., Tuan-Anh, P., Richard, J., Claeyssens, A., Carinos, E., de la Vieuville, G., & Matthee, J. 2023, Evolution of the Lyman-α-emitting fraction and UV properties of lensed star-forming galaxies in the range 2.9 ¡ z ¡ 6.7. *A&A*, 678, A174

Iani, E., Caputi, K. I., Rinaldi, P., Annunziatella, M., Boogaard, L. A., Östlin, G., Costantin, L., Gillman, S., Pérez-González, P. G., Colina, L., Greve, T. R., Wright, G., Alonso-Herrero, A., Álvarez-Márquez, J., Bik, A., Bosman, S. E. I., Crespo Gómez, A., Eckart, A., Hjorth, J., Jermann, I., Labiano, A., Langeroodi, D., Melinder, J., Moutard, T., Peißker, F., Pye, J. P., Tikkanen, T. V., van der Werf, P. P., Walter, F., Henning, T. K., Lagage, P.-O., & van Dishoeck, E. F. 2024, MIDIS: JWST NIRCam and MIRI Unveil the Stellar Population Properties of Lyα Emitters and Lyman-break Galaxies at $z \simeq 3$–7. *ApJ*, 963(2), 97

Jones, T., Stark, D. P., & Ellis, R. S. 2012, Keck Spectroscopy of Faint 3 ¡ z ¡ 7 Lyman Break Galaxies. III. The Mean Ultraviolet Spectrum at z ~= 4. *ApJ*, 751(1), 51

Orsi, A., Lacey, C. G., & Baugh, C. M. 2012, Can galactic outflows explain the properties of Ly α emitters? *MNRAS*, 425(1), 87–115

Ouchi, M., Shimasaku, K., Furusawa, H., Miyazaki, M., Doi, M., Hamabe, M., Hayashino, T., Kimura, M., Kodaira, K., Komiyama, Y., Matsuda, Y., Miyazaki, S., Nakata, F., Okamura, S., Sekiguchi, M., Shioya, Y., Tamura, H., Taniguchi, Y., Yagi, M., & Yasuda, N. 2003, Subaru Deep Survey. II. Luminosity Functions and Clustering Properties of Lyα Emitters at z=4.86 in the Subaru Deep Field. *ApJ*, 582(1), 60–68

Richard, J., Claeyssens, A., Lagattuta, D., Guaita, L., Bauer, F. E., Pello, R., Carton, D., Bacon, R., Soucail, G., Lyon, G. P., Kneib, J.-P., Mahler, G., Clément, B., Mercier, W., Variu, A., Tamone, A., Ebeling, H., Schmidt, K. B., Nanayakkara, T., Maseda, M., Weilbacher, P. M., Bouché, N., Bouwens, R. J., Wisotzki, L., de la Vieuville, G., Martinez, J., & Patrício, V. 2021, An atlas of MUSE observations towards twelve massive lensing clusters. *A&A*, 646, A83

Shapley, A. E., Steidel, C. C., Pettini, M., & Adelberger, K. L. 2003, Rest-Frame Ultraviolet Spectra of z~3 Lyman Break Galaxies. *ApJ*, 588(1), 65–89

Shibuya, T., Ouchi, M., Nakajima, K., Hashimoto, T., Ono, Y., Rauch, M., Gauthier, J.-R., Shimasaku, K., Goto, R., Mori, M., & Umemura., M. 2014, What is the Physical Origin of Strong Lyα Emission? II. Gas Kinematics and Distribution of Lyα Emitters. *ApJ*, 788(1), 74

Suess, K. A., Weaver, J. R., Price, S. H., Pan, R., Wang, B., Bezanson, R., Brammer, G., Cutler, S. E., Labbe, I., Leja, J., Williams, C. C., Whitaker, K. E., Dayal, P., de Graaff, A., Feldmann, R., Franx, M., Fudamoto, Y., Fujimoto, S., Furtak, L. J., Goulding, A. D., Greene, J. E., Khullar, G., Kokorev, V., Kriek, M., Lorenz, B., Marchesini, D., Maseda, M. V., Matthee, J., Miller, T. B., Mitsuhashi, I., Mowla, L. A., Muzzin, A., Naidu, R. P., Nanayakkara, T., Nelson, E. J., Oesch, P. A., Setton, D. J., Shipley, H., Smit, R., Spilker, J. S., van Dokkum, P., & Zitrin, A. 2024, Medium Bands, Mega Science: a JWST/NIRCam Medium-Band Imaging Survey of Abell 2744. *arXiv e-prints*, arXiv:2404.13132

Vanzella, E., Calura, F., Meneghetti, M., Mercurio, A., Castellano, M., Caminha, G. B., Balestra, I., Rosati, P., Tozzi, P., De Barros, S., Grazian, A., D'Ercole, A., Ciotti, L., Caputi, K.,

Grillo, C., Merlin, E., Pentericci, L., Fontana, A., Cristiani, S., & Coe, D. 2017, Paving the way for the JWST: witnessing globular cluster formation at z ¿ 3. *MNRAS*, 467(4), 4304–4321

Verhamme, A., Schaerer, D., & Maselli, A. 2006, 3D Lyα radiation transfer. I. Understanding Lyα line profile morphologies. *A&A*, 460(2), 397–413

Windhorst, R. A., Cohen, S. H., Jansen, R. A., Summers, J., Tompkins, S., Conselice, C. J., Driver, S. P., Yan, H., Coe, D., Frye, B., Grogin, N., Koekemoer, A., Marshall, M. A., O'Brien, R., Pirzkal, N., Robotham, A., Ryan, R. E., Willmer, C. N. A., Carleton, T., Diego, J. M., Keel, W. C., Porto, P., Redshaw, C., Scheller, S., Wilkins, S. M., Willner, S. P., Zitrin, A., Adams, N. J., Austin, D., Arendt, R. G., Beacom, J. F., Bhatawdekar, R. A., Bradley, L. D., Broadhurst, T., Cheng, C., Civano, F., Dai, L., Dole, H., D'Silva, J. C. J., Duncan, K. J., Fazio, G. G., Ferrami, G., Ferreira, L., Finkelstein, S. L., Furtak, L. J., Gim, H. B., Griffiths, A., Hammel, H. B., Harrington, K. C., Hathi, N. P., Holwerda, B. W., Honor, R., Huang, J.-S., Hyun, M., Im, M., Joshi, B. A., Kamieneski, P. S., Kelly, P., Larson, R. L., Li, J., Lim, J., Ma, Z., Maksym, P., Manzoni, G., Meena, A. K., Milam, S. N., Nonino, M., Pascale, M., Petric, A., Pierel, J. D. R., Polletta, M. d. C., Röttgering, H. J. A., Rutkowski, M. J., Smail, I., Straughn, A. N., Strolger, L.-G., Swirbul, A., Trussler, J. A. A., Wang, L., Welch, B., B. Wyithe, J. S., Yun, M., Zackrisson, E., Zhang, J., & Zhao, X. 2023, JWST PEARLS. Prime Extragalactic Areas for Reionization and Lensing Science: Project Overview and First Results. *AJ*, 165(1), 13

The First Chapters of Our Cosmic History with JWST
Proceedings IAU Symposium No. 391, 2025
T. Nanayakkara & M. Maseda, eds.
doi:10.1017/S1743921324002631

Primeval Lyman-α emitting galaxies reveal early sites of reionisation

Joris Witstok[1,2] & JADES Collaboration

[1]Cosmic Dawn Center (DAWN), Copenhagen, Denmark
email: joriswitstok@gmail.com
[2]Niels Bohr Institute, University of Copenhagen, Jagtvej 128, DK-2200, Copenhagen, Denmark

Abstract. Lyman-α (Lyα) emitting galaxies (LAEs) have been the subject of a range of observational studies of cosmic reionization, tracing the emergence of the first ionized 'bubbles' and the evolution of the neutral fraction of the intergalactic medium (IGM). Here we discuss a series of works characterising the physical properties and environments of the most distant LAEs identified in the JWST Advanced Deep Extragalactic Survey (JADES). Around half of the $z < 8$ LAEs fall within large-scale galaxy overdensities, providing strong evidence for enhanced Lyα transmission in these environments. Despite findings of strong Lyα absorption at the redshift frontier ($z > 9$) exceeding that expected only by the IGM, several LAEs are discovered at $z > 8$, including one at redshift $z = 13.0$ that shows highly exceptional properties. These observations firmly establish LAEs as powerful probes of the unfolding of cosmic reionisation out to the earliest stages of galaxy evolution.

Keywords. galaxies: high-redshift, dark ages, reionization, first stars, methods: observational, techniques: spectroscopic

1. Introduction

Within ΛCDM cosmology, the early growth of dark matter halos enables gas to collapse into the first structures, leading to the formation of the first galaxies at redshift $z \gtrsim 15$ (Dayal and Ferrara 2018). Hosting the first generation of stars, Population III (Pop III), they are thought to emit ionising photons and set in motion cosmic reionisation, while also quickly spreading heavy elements into the interstellar medium (ISM) through supernovae, thereby causing subsequent generations of stars to become slightly metal-enriched (Pop II; Bromm and Yoshida 2011). By $z \sim 5$, the (hydrogen) reionisation in the Universe had been fully completed (Bosman et al. 2022).

The atomic hydrogen (H I) Lyman-α (Lyα) transition has proven to be a powerful tracer not only of galaxies themselves, but also the reionisation process itself due to its sensitivity to being scattered by neutral gas in the intergalactic medium (IGM). Only once sufficiently large ionised regions have been formed, we may begin to observe Lyα-emitting galaxies (LAEs) within them, thus providing a key observable for characterising such early ionised 'bubbles' (Mason and Gronke 2020).

Indeed, large LAE surveys have shown that the fraction of galaxies observed as LAEs rapidly declines at $z > 6$, likely reflecting the evolving neutral fraction of the IGM (e.g. Stark et al. 2010; Pentericci et al. 2011). Meanwhile, at even earlier epochs ($z > 10$), signatures of very dense, neutral gas have been detected via strong Lyα damping-wing absorption ('damped Lyα' or DLA absorption; D'Eugenio et al. 2024b; Hsiao et al. 2023;

Hainline et al. 2024a), potentially indicating these galaxies are still in an early assembly phase (Heintz et al. 2024a,b). However, these damping-wing features may also be explained by dominant nebular continuum emission (Cameron et al. 2024; Katz et al. 2024), as also discussed in contributions by Kartheik Iyer and Michael Maseda.

In this contribution, we discuss the implications of a sample of distant LAEs observed as part of the JWST Advanced Deep Extragalactic Survey (JADES), a collaborative survey between NIRCam and NIRSpec GTO teams specifically aimed at charting galaxy evolution in the early Universe (Eisenstein et al. 2023).

2. The reionisation frontier: distant LAEs in the JADES survey

The JADES survey, detailed further in several data release papers (Rieke et al. 2023; Bunker et al. 2023a; D'Eugenio et al. 2024a), has obtained NIRSpec multi-object spectroscopy largely selected from NIRCam imaging data over the Great Observatories Origins Deep Survey (GOODS) fields (Giavalisco et al. 2004). The exposure times for the observations range from 10 to over 100 h per target across the resolution modes with $R \approx 100$ and $R \approx 1000$, allowing for unprecedented characterisation of faint, high-redshift galaxies. Since the first data was taken, many LAEs were newly discovered in JADES, as presented in Saxena et al. (2023, 2024) and Jones et al. (2024a,b).

In Witstok et al. (2024c), it was shown many of these LAEs coincide with large-scale galaxy overdensities identified with a large NIRCam wide-field slitless spectroscopy (WFSS) survey in the GOODS fields (FRESCO; Oesch et al. 2023), which is likely not a coincidence but rather a fundamental feature of ΛCDM cosmology, adding to a growing body of evidence that overdense regions become the first sites of reionisation (Jung et al. 2022, 2024; Chen et al. 2024; Witten et al. 2024), as discussed in the contribution by Nicolas Laporte.

With only two robust detections of $z > 8$ LAEs known prior to JWST (Zitrin et al. 2015; Larson et al. 2022), it was expected they would be very rare at the highest-redshift regime. It therefore came as a surprise that Lyα emission was clearly detected in GN-z11, a galaxy at $z \approx 10.6$ (Bunker et al. 2023b), even if it does show signs of being situated in an overdense environment (Tacchella et al. 2023). A further three LAEs at $8 < z < 9$ were presented by Witstok et al. (2024b). These LAEs are all compact, low-mass galaxies experiencing a recent burst of star formation (SF), causing SF surface densities well beyond those seen in local analogue galaxies. This is especially the case for the most extreme LAE observed, with a Lyα line appearing close to the systemic redshift at an equivalent width (EW) of ≈ 130 Å, suggesting their compactness likely aids the escape of both Lyα and ionising photons through stellar feedback processes. Nearby companion galaxies suggesting that mergers might play a role in triggering or enhancing star formation (cf. Witten et al. 2024).

Strong nebular emission lines were observed in both the rest-frame UV, where high-ionisation lines such as C IV suggest the presence of a hard radiation field (reaching >54 eV), and in the optical, where there is evidence of Balmer jumps. As in Witstok et al. (2024c), however, Witstok et al. (2024b) found these relatively faint LAEs ($M_{\rm UV} > -20$ mag or $L \lesssim 0.5 L^*$ at $z \approx 8$; Bouwens et al. 2021) are generally not able to inflate the large ionised bubbles they must inhabit to explain the observed Lyα emission, even when including their direct neighbours.

Intriguingly, deep NIRSpec follow-up spectroscopy presented by Witstok et al. (2024a) of an extremely compact, blue galaxy at $z = 13$ first identified by Hainline et al. (2024b) and Robertson et al. (2024) revealed remarkably strong Lyα emission (EW $\gtrsim 40$ Å). The spectrum reveals a clear continuum 'rollover', raising the question whether DLA absorption can be present at the same time as Lyα emission, or if this is instead due to the two-photon contribution to a dominant nebular continuum. The very blue UV continuum ($\beta_{\rm UV} \lesssim -2.7$), however, appears to indicate nebular emission is likely subdominant in the observed continuum emission. Indeed, this suggests a substantial fraction of ionising photons may be escaping the galaxy and creating an ionised bubble, only 330 Myr after the Big Bang. In any case, even if this galaxy is located inside an ionised bubble, the observed Lyα line strength suggests a highly efficient intrinsic ionising-photon production, $\xi_{\rm ion} \gtrsim 10^{26}\,{\rm Hz\,erg^{-1}}$.

3. Summary and Conclusions

The Lyα emission line continues to serve as a critical tool for studying the epoch of reionisation (EoR). The four $z > 8$ LAEs observed in JADES exhibit a range of Lyα properties, likely corresponding to different stages of local reionisation. All are compact, highly star-forming, low-mass galaxies. Finally, the discovery of an LAE at $z = 13$ pushes the boundaries of our understanding of reionisation, suggesting the presence of ionised regions in the very early Universe.

References

Bosman, S. E. I., Davies, F. B., Becker, G. D., Keating, L. C., Davies, R. L., Zhu, Y., Eilers, A.-C., D'Odorico, V., Bian, F., Bischetti, M., Cristiani, S. V., Fan, X., Farina, E. P., Haehnelt, M. G., Hennawi, J. F., Kulkarni, G., Mesinger, A., Meyer, R. A., Onoue, M., Pallottini, A., Qin, Y., Ryan-Weber, E., Schindler, J.-T., Walter, F., Wang, F., & Yang, J. 2022, Hydrogen reionization ends by z = 5.3: Lyman-α optical depth measured by the XQR-30 sample. *MNRAS*, 514(1), 55–76.

Bouwens, R. J., Oesch, P. A., Stefanon, M., Illingworth, G., Labbé, I., Reddy, N., Atek, H., Montes, M., Naidu, R., Nanayakkara, T., Nelson, E., & Wilkins, S. 2021, New Determinations of the UV Luminosity Functions from z 9 to 2 Show a Remarkable Consistency with Halo Growth and a Constant Star Formation Efficiency. *AJ*, 162(2), 47.

Bromm, V. & Yoshida, N. 2011, The First Galaxies. *ARA&A*, 49(1), 373–407.

Bunker, A. J., Cameron, A. J., Curtis-Lake, E., Jakobsen, P., Carniani, S., Curti, M., Witstok, J., Maiolino, R., D'Eugenio, F., Looser, T. J., Willott, C., Bonaventura, N., Hainline, K., Uebler, H., Willmer, C. N. A., Saxena, A., Smit, R., Alberts, S., Arribas, S., Baker, W. M., Baum, S., Bhatawdekar, R., Bowler, R. A. A., Boyett, K., Charlot, S., Chen, Z., Chevallard, J., Circosta, C., DeCoursey, C., de Graaff, A., Egami, E., Eisenstein, D. J., Endsley, R., Ferruit, P., Giardino, G., Hausen, R., Helton, J. M., Hviding, R. E., Ji, Z., Johnson, B. D., Jones, G. C., Kumari, N., Laseter, I., Luetzgendorf, N., Maseda, M. V., Nelson, E., Parlanti, E., Perna, M., Rauscher, B. J., Rawle, T., Rix, H.-W., Rieke, M., Robertson, B., Rodriguez Del Pino, B., Sandles, L., Scholtz, J., Sharpe, K., Skarbinski, M., Stark, D. P., Sun, F., Tacchella, S., Topping, M. W., Villanueva, N. C., Wallace, I. E. B., Williams, C. C., & Woodrum, C. 2023,a JADES NIRSpec Initial Data Release for the Hubble Ultra Deep Field: Redshifts and Line Fluxes of Distant Galaxies from the Deepest JWST Cycle 1 NIRSpec Multi-Object Spectroscopy. *arXiv e-prints*, arXiv:2306.02467.

Bunker, A. J., Saxena, A., Cameron, A. J., Willott, C. J., Curtis-Lake, E., Jakobsen, P., Carniani, S., Smit, R., Maiolino, R., Witstok, J., Curti, M., D'Eugenio, F., Jones, G. C., Ferruit, P., Arribas, S., Charlot, S., Chevallard, J., Giardino, G., de Graaff, A., Looser, T. J., Lützgendorf, N., Maseda, M. V., Rawle, T., Rix, H.-W., Del Pino, B. R., Alberts, S., Egami, E., Eisenstein, D. J., Endsley, R., Hainline, K., Hausen, R., Johnson, B. D., Rieke, G., Rieke, M., Robertson, B. E., Shivaei, I., Stark, D. P., Sun, F., Tacchella, S., Tang, M., Williams, C. C., Willmer, C. N. A., Baker, W. M., Baum, S., Bhatawdekar, R., Bowler, R., Boyett, K., Chen, Z., Circosta, C., Helton, J. M., Ji, Z., Kumari, N., Lyu, J., Nelson, E., Parlanti, E., Perna, M., Sandles, L., Scholtz, J., Suess, K. A., Topping, M. W., Übler, H., Wallace, I. E. B., & Whitler, L. 2023,b JADES NIRSpec Spectroscopy of GN-z11: Lyman-α

emission and possible enhanced nitrogen abundance in a z = 10.60 luminous galaxy. *A&A*, 677b, A88.

Cameron, A. J., Katz, H., Witten, C., Saxena, A., Laporte, N., & Bunker, A. J. 2024, Nebular dominated galaxies: insights into the stellar initial mass function at high redshift. *MNRAS*, 534(1), 523–543.

Chen, Z., Stark, D. P., Mason, C., Topping, M. W., Whitler, L., Tang, M., Endsley, R., & Charlot, S. 2024, JWST spectroscopy of z 5-8 UV-selected galaxies: new constraints on the evolution of the Ly α escape fraction in the reionization era. *MNRAS*, 528(4), 7052–7075.

Dayal, P. & Ferrara, A. 2018, Early galaxy formation and its large-scale effects. *Phys. Rep.*, 780, 1–64.

D'Eugenio, F., Cameron, A. J., Scholtz, J., Carniani, S., Willott, C. J., Curtis-Lake, E., Bunker, A. J., Parlanti, E., Maiolino, R., Willmer, C. N. A., Jakobsen, P., Robertson, B. E., Johnson, B. D., Tacchella, S., Cargile, P. A., Rawle, T., Arribas, S., Chevallard, J., Curti, M., Egami, E., Eisenstein, D. J., Kumari, N., Looser, T. J., Rieke, M. J., Rodríguez Del Pino, B., Saxena, A., Übler, H., Venturi, G., Witstok, J., Baker, W. M., Bhatawdekar, R., Bonaventura, N., Boyett, K., Charlot, S., Danhaive, A. L., Hainline, K. N., Hausen, R., Helton, J. M., Ji, X., Ji, Z., Jones, G. C., Joudžbalis, I., Maseda, M. V., Pérez-González, P. G., Perna, M., Puskás, D., Shivaei, I., Silcock, M. S., Simmonds, C., Smit, R., Sun, F., Villanueva, N. C., Williams, C. C., & Zhu, Y. 2024,a JADES Data Release 3 – NIRSpec/MSA spectroscopy for 4,000 galaxies in the GOODS fields. *arXiv e-prints*, arXiv:2404.06531.

D'Eugenio, F., Maiolino, R., Carniani, S., Chevallard, J., Curtis-Lake, E., Witstok, J., Charlot, S., Baker, W. M., Arribas, S., Boyett, K., Bunker, A. J., Curti, M., Eisenstein, D. J., Hainline, K., Ji, Z., Johnson, B. D., Kumari, N., Looser, T. J., Nakajima, K., Nelson, E., Rieke, M., Robertson, B., Scholtz, J., Smit, R., Sun, F., Venturi, G., Tacchella, S., Übler, H., Willmer, C. N. A., & Willott, C. 2024,b JADES: Carbon enrichment 350 Myr after the Big Bang. *A&A*, 689b, A152.

Eisenstein, D. J., Willott, C., Alberts, S., Arribas, S., Bonaventura, N., Bunker, A. J., Cameron, A. J., Carniani, S., Charlot, S., Curtis-Lake, E., D'Eugenio, F., Endsley, R., Ferruit, P., Giardino, G., Hainline, K., Hausen, R., Jakobsen, P., Johnson, B. D., Maiolino, R., Rieke, M., Rieke, G., Rix, H.-W., Robertson, B., Stark, D. P., Tacchella, S., Williams, C. C., Willmer, C. N. A., Baker, W. M., Baum, S., Bhatawdekar, R., Boyett, K., Chen, Z., Chevallard, J., Circosta, C., Curti, M., Danhaive, A. L., DeCoursey, C., de Graaff, A., Dressler, A., Egami, E., Helton, J. M., Hviding, R. E., Ji, Z., Jones, G. C., Kumari, N., Lützgendorf, N., Laseter, I., Looser, T. J., Lyu, J., Maseda, M. V., Nelson, E., Parlanti, E., Perna, M., Puskás, D., Rawle, T., Rodríguez Del Pino, B., Sandles, L., Saxena, A., Scholtz, J., Sharpe, K., Shivaei, I., Silcock, M. S., Simmonds, C., Skarbinski, M., Smit, R., Stone, M., Suess, K. A., Sun, F., Tang, M., Topping, M. W., Übler, H., Villanueva, N. C., Wallace, I. E. B., Whitler, L., Witstok, J., & Woodrum, C. 2023, Overview of the JWST Advanced Deep Extragalactic Survey (JADES). *arXiv e-prints*, arXiv:2306.02465.

Giavalisco, M., Ferguson, H. C., Koekemoer, A. M., Dickinson, M., Alexander, D. M., Bauer, F. E., Bergeron, J., Biagetti, C., Brandt, W. N., Casertano, S., Cesarsky, C., Chatzichristou, E., Conselice, C., Cristiani, S., Da Costa, L., Dahlen, T., de Mello, D., Eisenhardt, P., Erben, T., Fall, S. M., Fassnacht, C., Fosbury, R., Fruchter, A., Gardner, J. P., Grogin, N., Hook, R. N., Hornschemeier, A. E., Idzi, R., Jogee, S., Kretchmer, C., Laidler, V., Lee, K. S., Livio, M., Lucas, R., Madau, P., Mobasher, B., Moustakas, L. A., Nonino, M., Padovani, P., Papovich, C., Park, Y., Ravindranath, S., Renzini, A., Richardson, M., Riess, A., Rosati, P., Schirmer, M., Schreier, E., Somerville, R. S., Spinrad, H., Stern, D., Stiavelli, M., Strolger, L., Urry, C. M., Vandame, B., Williams, R., & Wolf, C. 2004, The Great Observatories Origins Deep Survey: Initial Results from Optical and Near-Infrared Imaging. *ApJ*, 600(2), L93–L98.

Hainline, K. N., D'Eugenio, F., Jakobsen, P., Chevallard, J., Carniani, S., Witstok, J., Ji, Z., Curtis-Lake, E., Johnson, B. D., Robertson, B., Tacchella, S., Curti, M., Charlot, S., Helton, J. M., Arribas, S., Bhatawdekar, R., Bunker, A. J., Cameron, A. J., Egami, E., Eisenstein,

D. J., Hausen, R., Kumari, N., Maiolino, R., Perez-Gonzalez, P. G., Rieke, M., Saxena, A., Scholtz, J., Smit, R., Sun, F., Williams, C. C., Willmer, C. N. A., & Willott, C. 2024,a Searching for Emission Lines at $z > 11$: The Role of Damped Lyman-α and Hints About the Escape of Ionizing Photons. *arXiv e-prints*, arXiv:2404.04325.

Hainline, K. N., Johnson, B. D., Robertson, B., Tacchella, S., Helton, J. M., Sun, F., Eisenstein, D. J., Simmonds, C., Topping, M. W., Whitler, L., Willmer, C. N. A., Rieke, M., Suess, K. A., Hviding, R. E., Cameron, A. J., Alberts, S., Baker, W. M., Baum, S., Bhatawdekar, R., Bonaventura, N., Boyett, K., Bunker, A. J., Carniani, S., Charlot, S., Chevallard, J., Chen, Z., Curti, M., Curtis-Lake, E., D'Eugenio, F., Egami, E., Endsley, R., Hausen, R., Ji, Z., Looser, T. J., Lyu, J., Maiolino, R., Nelson, E., Puskás, D., Rawle, T., Sandles, L., Saxena, A., Smit, R., Stark, D. P., Williams, C. C., Willott, C., & Witstok, J. 2024,b The Cosmos in Its Infancy: JADES Galaxy Candidates at z > 8 in GOODS-S and GOODS-N. *ApJ*, 964b(1), 71.

Heintz, K. E., Brammer, G. B., Watson, D., Oesch, P. A., Keating, L. C., Hayes, M. J., Abdurro'uf, Arellano-Córdova, K. Z., Carnall, A. C., Christiansen, C. R., Cullen, F., Davé, R., Dayal, P., Ferrara, A., Finlator, K., Fynbo, J. P. U., Flury, S. R., Gelli, V., Gillman, S., Gottumukkala, R., Gould, K., Greve, T. R., Hardin, S. E., Y. -Y Hsiao, T., Hutter, A., Jakobsson, P., Killi, M., Khosravaninezhad, N., Laursen, P., Lee, M. M., Magdis, G. E., Matthee, J., Naidu, R. P., Narayanan, D., Pollock, C., Prescott, M., Rusakov, V., Shuntov, M., Sneppen, A., Smit, R., Tanvir, N. R., Terp, C., Toft, S., Valentino, F., Vijayan, A. P., Weaver, J. R., Wise, J. H., & Witstok, J. 2024,a The JWST-PRIMAL Legacy Survey. A JWST/NIRSpec reference sample for the physical properties and Lyman-α absorption and emission of ~ 500 galaxies at $z = 5.5 - 13.4$. *arXiv e-prints*, arXiv:2404.02211.

Heintz, K. E., Watson, D., Brammer, G., Vejlgaard, S., Hutter, A., Strait, V. B., Matthee, J., Oesch, P. A., Jakobsson, P., Tanvir, N. R., Laursen, P., Naidu, R. P., Mason, C. A., Killi, M., Jung, I., Hsiao, T. Y.-Y., Abdurro'uf, Coe, D., Arrabal Haro, P., Finkelstein, S. L., & Toft, S. 2024,b Strong damped Lyman-α absorption in young star-forming galaxies at redshifts 9 to 11. *Science*, 384b(6698), 890–894.

Hsiao, T. Y.-Y., Abdurro'uf, Coe, D., Larson, R. L., Jung, I., Mingozzi, M., Dayal, P., Kumari, N., Kokorev, V., Vikaeus, A., Brammer, G., Furtak, L. J., Adamo, A., Andrade-Santos, F., Antwi-Danso, J., Bradac, M., Bradley, L. D., Broadhurst, T., Carnall, A. C., Conselice, C. J., Diego, J. M., Donahue, M., Eldridge, J. J., Fujimoto, S., Henry, A., Hernandez, S., Hutchison, T. A., James, B. L., Norman, C., Park, H., Pirzkal, N., Postman, M., Ricotti, M., Rigby, J. R., Vanzella, E., Welch, B., Wilkins, S. M., Windhorst, R. A., Xu, X., Zackrisson, E., & Zitrin, A. 2023, JWST NIRSpec spectroscopy of the triply-lensed $z = 10.17$ galaxy MACS0647−JD. *arXiv e-prints*, arXiv:2305.03042.

Jones, G. C., Bunker, A. J., Saxena, A., Arribas, S., Bhatawdekar, R., Boyett, K., Carniani, S., Charlot, S., Curtis-Lake, E., Hainline, K., Johnson, B. D., Kumari, N., Maseda, M. V., Rix, H.-W., Robertson, B. E., Tacchella, S., Übler, H., Williams, C. C., Willott, C., Witstok, J., & Zhu, Y. 2024,a JADES: Measuring reionization properties using Lyman-alpha emission. *arXiv e-prints*, arXiv:2409.06405.

Jones, G. C., Bunker, A. J., Saxena, A., Witstok, J., Stark, D. P., Arribas, S., Baker, W. M., Bhatawdekar, R., Bowler, R., Boyett, K., Cameron, A. J., Carniani, S., Charlot, S., Chevallard, J., Curti, M., Curtis-Lake, E., Eisenstein, D. J., Hainline, K., Hausen, R., Ji, Z., Johnson, B. D., Kumari, N., Looser, T. J., Maiolino, R., Maseda, M. V., Parlanti, E., Rix, H.-W., Robertson, B. E., Sandles, L., Scholtz, J., Smit, R., Tacchella, S., Übler, H., Williams, C. C., & Willott, C. 2024,b JADES: The emergence and evolution of Lyα emission and constraints on the intergalactic medium neutral fraction. *A&A*, 683b, A238.

Jung, I., Finkelstein, S. L., Arrabal Haro, P., Dickinson, M., Ferguson, H. C., Hutchison, T. A., Kartaltepe, J. S., Larson, R. L., Simons, R. C., Papovich, C., Park, H., Pentericci, L., Trump, J. R., Amorín, R. O., Backhaus, B. E., Bagley, M. B., Casey, C. M., Cheng, Y., Cleri, N. J., Cooper, M. C., Cooper, O. R., Gardner, J. P., Gawiser, E., Grazian, A., Hathi, N. P., Hirschmann, M., Koekemoer, A. M., Lucas, R. A., Mobasher, B., Pirzkal,

N., Ravindranath, S., Straughn, A. N., Yung, L. Y. A., & de la Vega, A. 2024, CEERS: Diversity of Lyα Emitters during the Epoch of Reionization. *ApJ*, 967(1), 73.

Jung, I., Papovich, C., Finkelstein, S. L., Simons, R. C., Estrada-Carpenter, V., Backhaus, B. E., Cleri, N. J., Finlator, K., Giavalisco, M., Ji, Z., Matharu, J., Momcheva, I., Straughn, A. N., & Trump, J. R. 2022, CLEAR: Boosted Lyα Transmission of the Intergalactic Medium in UV-bright Galaxies. *ApJ*, 933(1), 87.

Katz, H., Cameron, A. J., Saxena, A., Barrufet, L., Choustikov, N., Cleri, N. J., de Graaff, A., Ellis, R. S., Fosbury, R. A. E., Heintz, K. E., Maseda, M., Matthee, J., McConchie, I., & Oesch, P. A. 2024, 21 Balmer Jump Street: The Nebular Continuum at High Redshift and Implications for the Bright Galaxy Problem, UV Continuum Slopes, and Early Stellar Populations. *arXiv e-prints*, arXiv:2408.03189.

Larson, R. L., Finkelstein, S. L., Hutchison, T. A., Papovich, C., Bagley, M., Dickinson, M., Rojas-Ruiz, S., Ferguson, H. C., Jung, I., Giavalisco, M., Grazian, A., Pentericci, L., & Tacchella, S. 2022, Searching for Islands of Reionization: A Potential Ionized Bubble Powered by a Spectroscopic Overdensity at z = 8.7. *ApJ*, 930(2), 104.

Mason, C. A. & Gronke, M. 2020, Measuring the properties of reionized bubbles with resolved Lyα spectra. *MNRAS*, 499(1), 1395–1405.

Oesch, P. A., Brammer, G., Naidu, R. P., Bouwens, R. J., Chisholm, J., Illingworth, G. D., Matthee, J., Nelson, E., Qin, Y., Reddy, N., Shapley, A., Shivaei, I., van Dokkum, P., Weibel, A., Whitaker, K., Wuyts, S., Covelo-Paz, A., Endsley, R., Fudamoto, Y., Giovinazzo, E., Herard-Demanche, T., Kerutt, J., Kramarenko, I., Labbe, I., Leonova, E., Lin, J., Magee, D., Marchesini, D., Maseda, M., Mason, C., Matharu, J., Meyer, R. A., Neufeld, C., Prieto Lyon, G., Schaerer, D., Sharma, R., Shuntov, M., Smit, R., Stefanon, M., Wyithe, J. S. B., & Xiao, M. 2023, The JWST FRESCO survey: legacy NIRCam/grism spectroscopy and imaging in the two GOODS fields. *MNRAS*, 525(2), 2864–2874.

Pentericci, L., Fontana, A., Vanzella, E., Castellano, M., Grazian, A., Dijkstra, M., Boutsia, K., Cristiani, S., Dickinson, M., Giallongo, E., Giavalisco, M., Maiolino, R., Moorwood, A., Paris, D., & Santini, P. 2011, Spectroscopic Confirmation of z ~7 Lyman Break Galaxies: Probing the Earliest Galaxies and the Epoch of Reionization. *ApJ*, 743(2), 132.

Rieke, M. J., Robertson, B., Tacchella, S., Hainline, K., Johnson, B. D., Hausen, R., Ji, Z., Willmer, C. N. A., Eisenstein, D. J., Puskás, D., Alberts, S., Arribas, S., Baker, W. M., Baum, S., Bhatawdekar, R., Bonaventura, N., Boyett, K., Bunker, A. J., Cameron, A. J., Carniani, S., Charlot, S., Chevallard, J., Chen, Z., Curti, M., Curtis-Lake, E., Danhaive, A. L., DeCoursey, C., Dressler, A., Egami, E., Endsley, R., Helton, J. M., Hviding, R. E., Kumari, N., Looser, T. J., Lyu, J., Maiolino, R., Maseda, M. V., Nelson, E. J., Rieke, G., Rix, H.-W., Sandles, L., Saxena, A., Sharpe, K., Shivaei, I., Skarbinski, M., Smit, R., Stark, D. P., Stone, M., Suess, K. A., Sun, F., Topping, M., Übler, H., Villanueva, N. C., Wallace, I. E. B., Williams, C. C., Willott, C., Whitler, L., Witstok, J., & Woodrum, C. 2023, JADES Initial Data Release for the Hubble Ultra Deep Field: Revealing the Faint Infrared Sky with Deep JWST NIRCam Imaging. *ApJS*, 269(1), 16.

Robertson, B., Johnson, B. D., Tacchella, S., Eisenstein, D. J., Hainline, K., Arribas, S., Baker, W. M., Bunker, A. J., Carniani, S., Cargile, P. A., Carreira, C., Charlot, S., Chevallard, J., Curti, M., Curtis-Lake, E., D'Eugenio, F., Egami, E., Hausen, R., Helton, J. M., Jakobsen, P., Ji, Z., Jones, G. C., Maiolino, R., Maseda, M. V., Nelson, E., Pérez-González, P. G., Puskás, D., Rieke, M., Smit, R., Sun, F., Übler, H., Whitler, L., Williams, C. C., Willmer, C. N. A., Willott, C., & Witstok, J. 2024, Earliest Galaxies in the JADES Origins Field: Luminosity Function and Cosmic Star Formation Rate Density 300 Myr after the Big Bang. *ApJ*, 970(1), 31.

Saxena, A., Bunker, A. J., Jones, G. C., Stark, D. P., Cameron, A. J., Witstok, J., Arribas, S., Baker, W. M., Baum, S., Bhatawdekar, R., Bowler, R., Boyett, K., Carniani, S., Charlot, S., Chevallard, J., Curti, M., Curtis-Lake, E., Eisenstein, D. J., Endsley, R., Hainline, K., Helton, J. M., Johnson, B. D., Kumari, N., Looser, T. J., Maiolino, R., Rieke, M., Rix, H.-W., Robertson, B. E., Sandles, L., Simmonds, C., Smit, R., Tacchella, S., Williams, C. C.,

Willmer, C. N. A., & Willott, C. 2024, JADES: The production and escape of ionizing photons from faint Lyman-alpha emitters in the epoch of reionization. *A&A*, 684, A84.

Saxena, A., Robertson, B. E., Bunker, A. J., Endsley, R., Cameron, A. J., Charlot, S., Simmonds, C., Tacchella, S., Witstok, J., Willott, C., Carniani, S., Curtis-Lake, E., Ferruit, P., Jakobsen, P., Arribas, S., Chevallard, J., Curti, M., D'Eugenio, F., De Graaff, A., Jones, G. C., Looser, T. J., Maseda, M. V., Rawle, T., Rix, H.-W., Del Pino, B. R., Smit, R., Übler, H., Eisenstein, D. J., Hainline, K., Hausen, R., Johnson, B. D., Rieke, M., Williams, C. C., Willmer, C. N. A., Baker, W. M., Bhatawdekar, R., Bowler, R., Boyett, K., Chen, Z., Egami, E., Ji, Z., Kumari, N., Nelson, E., Perna, M., Sandles, L., Scholtz, J., & Shivaei, I. 2023, JADES: Discovery of extremely high equivalent width Lyman-α emission from a faint galaxy within an ionized bubble at z = 7.3. *A&A*, 678, A68.

Stark, D. P., Ellis, R. S., Chiu, K., Ouchi, M., & Bunker, A. 2010, Keck spectroscopy of faint 3 < z < 7 Lyman break galaxies - I. New constraints on cosmic reionization from the luminosity and redshift-dependent fraction of Lyman α emission. *MNRAS*, 408(3), 1628–1648.

Tacchella, S., Eisenstein, D. J., Hainline, K., Johnson, B. D., Baker, W. M., Helton, J. M., Robertson, B., Suess, K. A., Chen, Z., Nelson, E., Puskás, D., Sun, F., Alberts, S., Egami, E., Hausen, R., Rieke, G., Rieke, M., Shivaei, I., Williams, C. C., Willmer, C. N. A., Bunker, A., Cameron, A. J., Carniani, S., Charlot, S., Curti, M., Curtis-Lake, E., Looser, T. J., Maiolino, R., Maseda, M. V., Rawle, T., Rix, H.-W., Smit, R., Übler, H., Willott, C., Witstok, J., Baum, S., Bhatawdekar, R., Boyett, K., Danhaive, A. L., de Graaff, A., Endsley, R., Ji, Z., Lyu, J., Sandles, L., Saxena, A., Scholtz, J., Topping, M. W., & Whitler, L. 2023, JADES Imaging of GN-z11: Revealing the Morphology and Environment of a Luminous Galaxy 430 Myr after the Big Bang. *ApJ*, 952(1), 74.

Witstok, J., Jakobsen, P., Maiolino, R., Helton, J. M., Johnson, B. D., Robertson, B. E., Tacchella, S., Cameron, A. J., Smit, R., Bunker, A. J., Saxena, A., Sun, F., Arribas, S., Baker, W. M., Bhatawdekar, R., Boyett, K., Cargile, P. A., Carniani, S., Charlot, S., Chevallard, J., Curti, M., Curtis-Lake, E., D'Eugenio, F., Eisenstein, D. J., Hainline, K. N., Jones, G. C., Kumari, N., Maseda, M. V., Pérez-González, P. G., Rinaldi, P., Scholtz, J., Übler, H., Williams, C. C., Willmer, C. N. A., Willott, C., & Zhu, Y. 2024,a Witnessing the onset of Reionisation via Lyman-α emission at redshift 13. *arXiv e-prints*, arXiv:2408.16608.

Witstok, J., Maiolino, R., Smit, R., Jones, G. C., Bunker, A. J., Helton, J. M., Johnson, B. D., Tacchella, S., Saxena, A., Arribas, S., Bhatawdekar, R., Boyett, K., Cameron, A. J., Cargile, P. A., Carniani, S., Charlot, S., Chevallard, J., Curti, M., Curtis-Lake, E., D'Eugenio, F., Eisenstein, D. J., Hainline, K., Hausen, R., Kumari, N., Laseter, I., Maseda, M. V., Rieke, M., Robertson, B., Scholtz, J., Shivaei, I., Williams, C. C., Willmer, C. N. A., & Willott, C. 2024,b JADES: Primeval Lyman-α emitting galaxies reveal early sites of reionisation out to redshift $z \sim 9$. *arXiv e-prints*, arXiv:2404.05724.

Witstok, J., Smit, R., Saxena, A., Jones, G. C., Helton, J. M., Sun, F., Maiolino, R., Kumari, N., Stark, D. P., Bunker, A. J., Arribas, S., Baker, W. M., Bhatawdekar, R., Boyett, K., Cameron, A. J., Carniani, S., Charlot, S., Chevallard, J., Curti, M., Curtis-Lake, E., Eisenstein, D. J., Endsley, R., Hainline, K., Ji, Z., Johnson, B. D., Looser, T. J., Nelson, E., Perna, M., Rix, H.-W., Robertson, B. E., Sandles, L., Scholtz, J., Simmonds, C., Tacchella, S., Übler, H., Williams, C. C., Willmer, C. N. A., & Willott, C. 2024,c Inside the bubble: exploring the environments of reionisation-era Lyman-α emitting galaxies with JADES and FRESCO. *A&A*, 682c, A40.

Witten, C., Laporte, N., Martin-Alvarez, S., Sijacki, D., Yuan, Y., Haehnelt, M. G., Baker, W. M., Dunlop, J. S., Ellis, R. S., Grogin, N. A., Illingworth, G., Katz, H., Koekemoer, A. M., Magee, D., Maiolino, R., McClymont, W., Pérez-González, P. G., Puskás, D., Roberts-Borsani, G., Santini, P., & Simmonds, C. 2024, Deciphering Lyman-α emission deep into the epoch of reionization. *Nature Astronomy*, 8, 384–396.

Zitrin, A., Labbé, I., Belli, S., Bouwens, R., Ellis, R. S., Roberts-Borsani, G., Stark, D. P., Oesch, P. A., & Smit, R. 2015, Lymanα Emission from a Luminous z = 8.68 Galaxy: Implications for Galaxies as Tracers of Cosmic Reionization. *ApJ*, 810(1), L12.

The First Chapters of Our Cosmic History with JWST
Proceedings IAU Symposium No. 391, 2025
T. Nanayakkara & M. Maseda, eds.
doi:10.1017/S1743921324002539

Very massive stars and Nitrogen-emitting galaxies

Jorick S. Vink

Armagh Observatory and Planetarium
email: jorick.vink@armagh.ac.uk

Abstract. Recent studies of high-redshift galaxies using JWST, such as GN-z11 revealed highly elevated levels of nitrogen (N). This phenomenon extends to gravitationally-lensed galaxies like the Sunburst Arc at z = 2.37, as well as to globular clusters (GCs). We propose that this originates from the presence of very massive stars (VMSs) with masses ranging from 100 to 1000 $M_\odot$. The He II observed in the Sunburst Arc could also stem from the disproportionately large contribution of VMSs. We build an entirely new Framework for massive star evolution which is no longer set by Dutch or other mass-loss "recipes" but which take the physics of Γ or L/M-dependent winds into account. We discuss the mass-loss kink and the transition mass-loss rate between optically thin and thick winds, before we study the evaporative mass-loss history of VMSs. Our novel evolution models exhibit vertical evolution in the HR-diagram from the zero-age main sequence due to a self-regulatory effect driven by their wind-dominated nature, and we discuss what wind physics sets the stellar upper-mass limit. Our estimate for the Sunburst Arc in Vink (2023) suggests that the significant amounts of N found in star-forming galaxies likely arise from VMSs. We evaluate the strengths and weaknesses of previous hypotheses, including fast rotating massive stars and supermassive stars (SMSs), and we conclude that only our VMS model satisfies the relevant criteria. Finally, we advocate for the inclusion of VMSs in population synthesis and chemical evolution models, emphasizing the need for a self-consistent wind approach, which currently does not exist. Even minor inaccuracies in mass-loss rates dramatically impact the stellar evolution of VMS, as well as their ionizing and chemical feedback.

Keywords. Star formation, yields, massive stars, stellar winds

1. Introduction

A central question in Astrophysics is whether there is an upper mass limit for stars and, if so, what physical mechanism determines this limit. The answer to this question could also indicate whether the upper mass limit depends on metallicity (Z). We proposed that mass loss due to radiation-driven winds influenced by line opacity is a key factor in determining the stellar upper mass limit (Vink 2018).

High-redshift galaxies are now commonly investigated using JWST. One surprising finding from galaxies like GN-z11 are the elevated nitrogen (N) abundances (Cameron et al. 2023; Senchyna et al. 2023; Marques-Chaves et al. 2023; Pascale et al. 2023). A plausible explanation for this N enhancement is the production and subsequent stellar wind release of hydrogen (H) burning products through the CNO cycle.

Very massive stars (VMS) with masses in the range of 100–1000 $M_\odot$ are formidable cosmic engines, as well as factories of chemical elements, such as N (Vink 2023), as well as sodium (Na), and aluminium (Al) relevant for globular clusters (Vink 2018; Higgins et al. 2023).

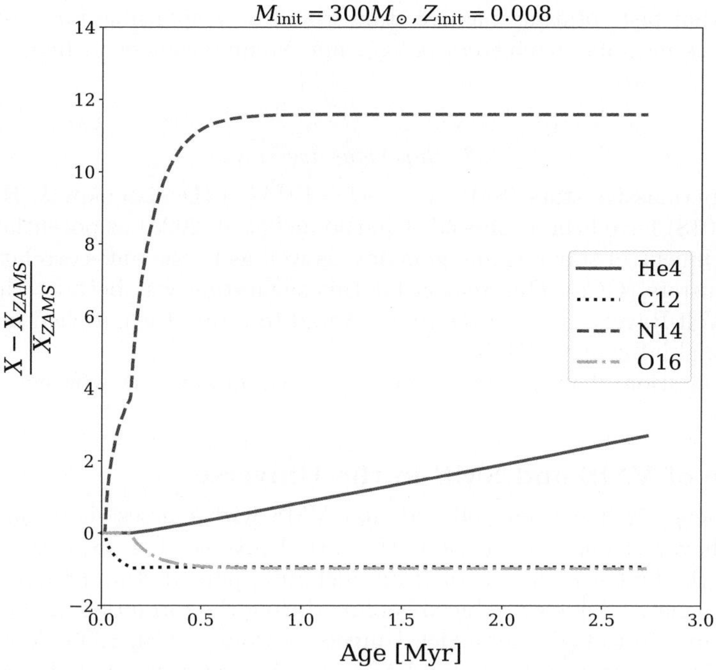

Figure 1. Relative surface enrichment for a 300 $M_\odot$ VMS at sub-solar LMC-like metallicity. The VMS MESA stellar evolution models are from Sabhahit et al. (2022).

What makes VMS with masses in the 100–1000$M_\odot$ range fundamentally different from canonical O stars in 20–100 $M_\odot$ range is their proximity to the Eddington Γ limit (Figure 1). VMS have huge convective cores, and with enhanced mass-loss rates (Vink et al. 2011) they naturally undergo chemical homogeneous evolution (CHE) (Gräfener et al. 2011; Yusof et al. 2013; Sabhahit et al. 2022) without the need to invoke CHE from fast stellar rotation (Vink & Harries 2017).

2. Alternative Polluters

2.1. *Asymptotic Giant Branch stars*

While AGB stars are key producers of chemical elements in the Cosmos, it is unlikely that they are responsible for chemical enrichment of the Early Universe, as it takes a long evolutionary time for low and intermediate mass-stars to reach the AGB phase. In addition, the anti-correlations in GCs between Na and oxygen are not reproduced (Bastian & Lardo 2018) For these reasons, massive stars appear to be the more likely culprits, but which massive stars?

2.2. *Fast rotating massive stars*

Another class of stellar objects that have been suggested to reproduce the anti-correlations in GCs are fast rotating massive stars (Decressin et al. 2007). However, this scenario would rely on a number of assumptions, such as high stellar rotation, end very efficient rotational mixing. Rotating stars leading to wind ejection in classical Wolf-Rayet (cWR) stars have also been invoked to explain the N in high-z galaxies (Senchyna et al. 2023), but it should be realised that if cWR stars at lower Z would indeed rotate more rapidly, the expectation would be that they should be more linearly polarised than their Galactic cWR counterparts, which is not observed (Vink & Harries 2017). It is thus questionable that fast rotating massive stars would be responsible for Na-O and Al-Mg anti-correlations in globular clusters (GCs), nor N-enhancement in high z star-forming galaxies.

2.3. *Supermassive stars*

Instead, supermassive stars (SMS, with $M \simeq 10^4 M_\odot$) (Denissenkov & Hartwick 2014; Gieles et al. 2018) have been proposed (Charbonnel et al. 2023) as potential contributors to the N enrichment in star-forming galaxies, as well as to the anti-correlations observed in globular clusters (GCs). One reason for this suggestion was because the winds from carbon-rich Wolf-Rayet stars (cWR) are too rapid to retain the enriched material within the gravitational influence of GCs.

A critical question that has not been addressed however is whether SMS exist in Nature.

3. The role of VMS and SMS in the Universe

For many years, it had been believed that VMS with masses significantly exceeding $100\,M_\odot$ may have been more common in the early Universe. This expectation stems from the idea that the first generations of stars, including pop III stars (Schaerer 2002) were generally more massive due to reduced cooling during the formation of these metal-poor objects compared to today's metal-rich Universe (Bromm et al. 1999; Abel et al. 2000).

But there is a second reason why VMS may be more common in the metal-poor Universe that that is due to the physics of radiation-driven winds that are thought to be weaker in the early Universe with its lower metal content, leading to a higher upper mass limit at lower Z (Vink 2018). This could imply that the final masses of VMS would remain close to their initial masses, which could even lead to the creation of intermediate-mass black holes from VMS in the 100–1000 $M_\odot$ range, or even the direct creation of supermassive black holes around 10^5 $M_\odot$ found in galactic centers, if SMS exist.

4. The very existence of VMS

Prior to addressing the reality of SMS we should first recall the evidence for the existence of VMS. Crowther et al. (2010) conducted a re-analysis of the most massive hydrogen- and nitrogen-rich Wolf-Rayet (WNh) stars located in the center of R136, the ionizing cluster of the Tarantula Nebula in the Large Magellanic Cloud (LMC). Their analysis revealed that stars previously thought to be below a "canonical" upper mass limit of $150 M_\odot$ were actually more luminous, with initial masses reaching approximately 200-300$M_\odot$ (Vink 2015; Martins & Palacios 2022).

Some members of the astronomical community appear to remain skeptical about the existence of very high masses in R136, particularly in light of earlier extraordinary claims about SMS of thousands of solar masses in the Tarantula nebula in the 1980s, as higher-resolution observations have revealed that R136 was not a single SMS but rather a young cluster containing several lower-mass objects, including the current record holder, R136a1.

For some it thus appears to remain questionable that the luminosities derived by Crowther et al. are overestimates, as the central WNh stars might actually involve multiple sources due to inadequate spatial resolution.

However, while it would technically be possible that the most luminous stars in areas such as R136 could dissolve into smaller objects from a photometry point-of-view alone, the reality is that these WNh stars have very specific emission line spectra that can only be produced by VMS with extremely strong stellar winds (Vink et al. 2011). By contrast, canonical O-type stars are predominately characterised by absorption lines.

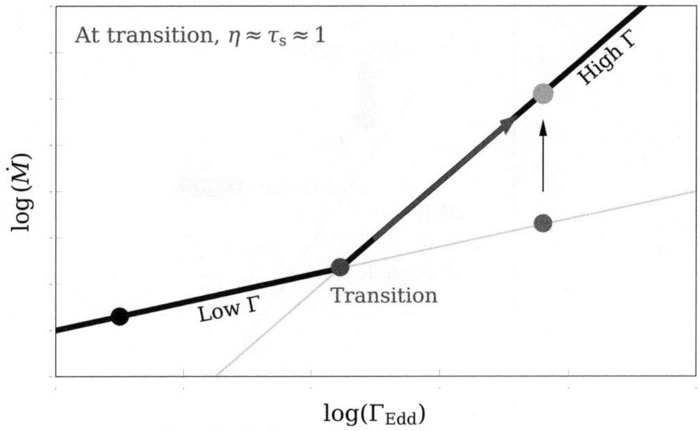

Figure 2. A cartoon of the wind mass-loss rate versus the Eddington parameter Γ. The black dot represents stars in the traditional massive O-type regime, while the red dot provides the Vink & Gräfener (2012) transition point. Above this red point, wind mass-loss rates need to be boosted from the blue to green dot, accounting for the much steeper Vink et al. (2011) slope at higher Γ (see Sabhahit et al. (2022).

As the VMS in R136 are in the emission-line dominated regime, breaking up one $300M_{\odot}$ star into say ten $30M_{\odot}$ O-type stars cannot work, as that would lead to an integrated absorption line spectrum for R136, while He II is in fact observed in emission (Crowther et al. 2016). Let us examine the transition from optically thin ($\tau < 1$) O-type stars to optically thick ($\tau > 1$) VMS in the following.

5. The extreme winds of VMS

Canonical O-type stars stellar spectra that are characterised by absorption lines, but when stars become more luminous, and display increasingly larger mass-loss rates, He II absorption lines turn into emission (Vink et al. 2011; Bestenlehner et al. 2014). This change-over happens when the wind optical depth τ and the wind efficiency number η cross unity (Vink & Gräfener 2012).

Figure 2 shows a kink in the relationship between mass-loss rate and the Eddington Γ parameter (which is proportional to the L/M ratio). The O-type stars on the lower slope of the relation have absorption lines, whilst those above the Vink & Gräfener (2012) $\eta = \tau = 1$ mass-loss transition "kink" have emission lines.

In addition to affecting the integrated spectra of star-forming galaxies (Leitherer et al. 1999; Meštrić et al. 2023), the stronger mass-loss slope of VMS also has a decisive effect on the stellar evolution, the final black hole (BH) mass, and the chemical yields from strong stellar winds.

For this reason, in Armagh we have recently implemented the $\eta = \tau = 1$ mass-loss transition strategy into the MESA (Paxton et al. 2013) stellar evolution code (Sabhahit et al. 2022, 2023), and we encounter vertical evolution in the HR diagram, as opposed to traditional horizontal evolution (see Figure. 3). The enhanced wind mass loss also leads to a mass-evaporation effects, with consequences for the upper mass limit to stars (Vink 2018).

6. Mass Evaporation and the Hydrogen Clock

When we include the enhanced winds into stellar evolution models, we obtain a strong drop of the stellar mass with time, as shown in Figure. 4. In particular, a convergence to a similar stellar mass is observed around 1.5 million years – regardless of wether the initial mass was say $300\,M_{\odot}$ or $1000\,M_{\odot}$.

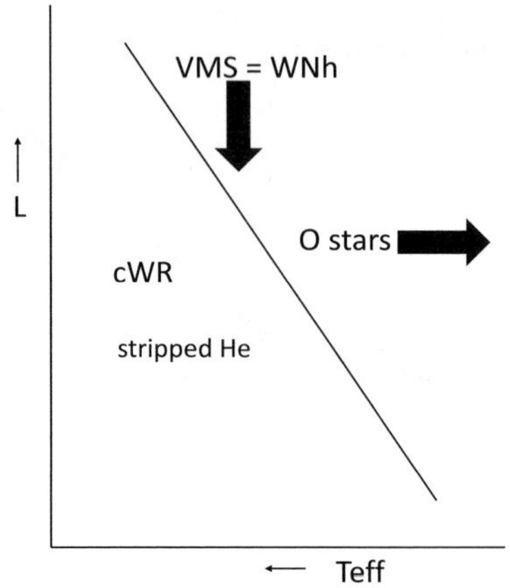

Figure 3. While canonical massive O-type stars are expected to evolve horizontally in the stellar HR diagram (HRD), very massive stars (VMS) are observed to evolve vertically downwards from the onset of hydrogen burning, as indicated by the downward thick arrow. These stars are likely to produce classical Wolf-Rayet stars (cWRs) located on the hot side of the slanted line representing the zero-age main sequence (ZAMS). At luminosities below a certain threshold (Sander & Vink 2020), these stars may appear as optically thin stripped helium stars.

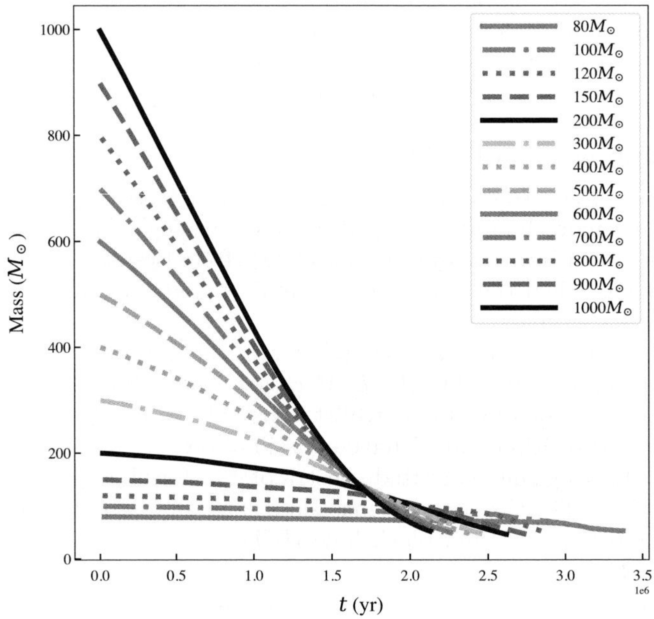

Figure 4. The "effective upper mass limit" and stellar mass evolution of VMS up to 1000 $M_\odot$, as they progress from Zero Age Main Sequence (ZAMS) to Terminal Age Main Sequence (TAMS). The alternating gray lines indicate the different masses within the grid. The figure emphasizes that it is not feasible to determine the initial upper mass limit of stars based on present-day observations, unless one knows the precise mass-loss history (Vink 2018; Sabhahit et al. 2022; Higgins et al. 2022).

Furthermore, the surface hydrogen abundance X_S can serve as the key for determining initial mass, as VMS evolve chemically homogeneous above a certain mass threshold. In Higgins et al. (2022) we utilized this tool to resolve the ambiguity in the initial masses of both components in a detached binary system as well as amongst a larger sample of

WNh stars in the Tarantula Nebula. For some objects we found that the initial stellar mass is completely unrestricted, with options easily extending to $1000M_\odot$.

In short, the mass turnover point at 1.5 million years in combination with X_S can function as a 'clock' for estimating the stellar upper mass limit.

7. Summary

• VMS in the 100–1000 $M_\odot$ range are *fundamentally* different from canonical O stars in the 20–100$M_\odot$ range. For this reason, simply discussing a potentially "top-heavy" nature of the IMF of high-redshift star forming galaxies has very limited value.

• VMS produce the He II emission from young stellar clusters in the Local Universe.

• VMS may produce excessive amounts of N, and could be responsible for the exciting JWST observations of galaxies such as GN z-11. Additional "by-products" from H-burning via the CNO cycle, such as Na and Al, could be relevant for the self-enrichment of globular clusters.

References

Abel, T., Bryan, G. L., & Norman, M. L. 2000, *ApJ* 540, 39
Bastian, N. & Lardo, C. 2018, *ARA&A* 56, 83
Bestenlehner, J. M., Gräfener, G., Vink, J. S., et al. 2014, *A&A* 570, A38
Bromm, V., Coppi, P. S., & Larson, R. B. 1999, *ApJL* 527, L5
Cameron, A. J., Katz, H., Rey, M. P., et al. 2023, *MNRAS* 523, 3516
Charbonnel, C., Schaerer, D., Prantzos, N., et al. 2023, *A&A* 673, L7
Crowther P. A., Schnurr O., Hirschi R., et al., 2010, *MNRAS*, 408, 731
Crowther, P. A., Caballero-Nieves, S. M., Bostroem, K. A., et al. 2016, *MNRAS* 458, 624
Decressin, T., Meynet, G., Charbonnel, C., Prantzos, N., & Ekström, S. 2007, *A&A* 464, 1029
Denissenkov, P. A., & Hartwick, F. D. A. 2014, *MNRAS* 437, L21
Gieles, M., Charbonnel, C., Krause, M. G. H., et al. 2018, *MNRAS* 478, 2461
Gräfener, G., Vink, J. S., de Koter, A., et al. 2011, *A&A* 535, A56
Higgins, E. R., Vink, J. S., Sabhahit, G. N., et al. 2022, *MNRAS* 516, 4052
Higgins, E. R., Vink, J. S., Hirschi, R., et al. 2023, *MNRAS* 526, 534
Leitherer, C., Schaerer, D., Goldader, J. D., et al. 1999, *ApJS* 123, 3
Marques-Chaves, R., Schaerer, D., Kuruvanthodi, A., et al. 2023, arXiv:2307.04234
Martins, F. & Palacios, A. 2022, *A&A* 659, A163
Meštrić, U., Vanzella, E., Upadhyaya, A., et al. 2023, *A&A* 673, A50
Pascale, M., Dai, L., McKee, C. F., et al. 2023, arXiv:2301.10790.
Paxton B., Cantiello M., Arras P., et al., 2013, *ApJS*, 208, 4
Sabhahit, G. N., Vink, J. S., Higgins, E. R., et al. 2022, *MNRAS* 514, 3736
Sabhahit, G. N., Vink, J. S., Sander, A. A. C., et al. 2023, *MNRAS* 524, 1529
Sander, A. A. C. & Vink, J. S. 2020, *MNRAS* 499, 873
Schaerer, D. 2002, *A&A* 382, 28
Senchyna, P., Plat, A., Stark, D. P., et al. 2023, arXiv:2303.04179
Vink, J. S. 2015, Very Massive Stars in the Local Universe 412, arXiv:1406.4836.
Vink, J. S. 2018, *A&A* 615, A119
Vink, J. S. 2023, *A&A* 679, L9
Vink, J. S., & Gräfener, G., 2012, *ApJL* 751, 34
Vink, J. S. & Harries, T. J. 2017, *A&A* 603, A120
Vink, J. S., Muijres, L. E., Anthonisse, B., et al., 2011, *A&A* 531, 132
Yusof, N., Hirschi, R., Meynet, G., et al. 2013, *MNRAS* 433, 1114

The First Chapters of Our Cosmic History with JWST
Proceedings IAU Symposium No. 391, 2025
T. Nanayakkara & M. Maseda, eds.
doi:10.1017/S174392132400276X

Stacking high-z galaxies' spectra from JWST/NIRSpec to explore the connection between Lyman alpha emission (or absence thereof) and rest-frame UV and optical properties

Nimisha Kumari on behalf of JADES collaboration

AURA for European Space Agency (ESA), ESA Office, Space Telescope Science Institute, 3700 San Martin Drive, Baltimore, MD, 21218, USA
email: kumari@stsci.edu

Abstract. Understanding the epoch of reionization (EoR) is one of the frontier goals of modern astronomy. Lyα emission from the high-redshift galaxies has proved to be a useful tool for probing cosmic reionization. The launch of JWST has enabled the ultraviolet and optical spectroscopic studies of Lyα emitters (LAE) in the EoR. In this contribution, we present the results based on the stacking analysis of JWST/NIRSpec spectra of galaxies taken as part of the JWST GTO program JADES. The sample consists of >250 galaxies at redshifts of $\sim$5-10. We divide this sample into six different sub-samples of galaxies based on the strengths of Lyα emission and redshifts and create composite spectra. We then estimate emission line ratios of optical and UV emission lines, and various physical quantities such as dust extinction, electron temperatures, ionization parameters, and ionizing photon production efficiencies to characterize the populations of LAEs and non-LAEs.

Keywords. Lyman alpha, Reionization era, JWST/NIRSpec

1. Introduction

One of the enduring challenges of astronomy is to develop a clear and lucid understanding of the history of the Universe. As per the Lamda Cold Dark Model, the Big Bang happened about 13.8 billion years ago, ensuing the recombination era and cosmic dark ages, followed by the formation of the first stars and galaxies, leading to the reionization era (Figure 1).

There are several questions revolving around the epoch of reionization, for example, how and when the process of reionization happened; what were the potential sources of reionization; were they active galactic nuclei, quasars, early star-forming galaxies, Lyman break galaxies, Lyman alpha emitters (LAEs); what were the distinctive properties or features of these sources?

Since the seminal work of Partridge & Peebles (1967) suggesting the association of Lyα emission with the early galaxies, several efforts are made to search for Lyα emitters. The launch of James Webb Space Telescope (JWST) has greatly improved the scenario, with Near-Infrared Spectrograph (NIRSpec) leading not only to the Lyα detection out to redshifts z$\sim$11 (e.g., Bunker et al. 2023), but also enabling the detailed spectroscopic studies of high-redshift Lyα emitting galaxies (e.g., Saxena *et al.* 2023; Jones *et al.* 2024; Kumari *et al.* 2024; Roberts-Borsani *et al.* 2024; Witstok *et al.* 2024).

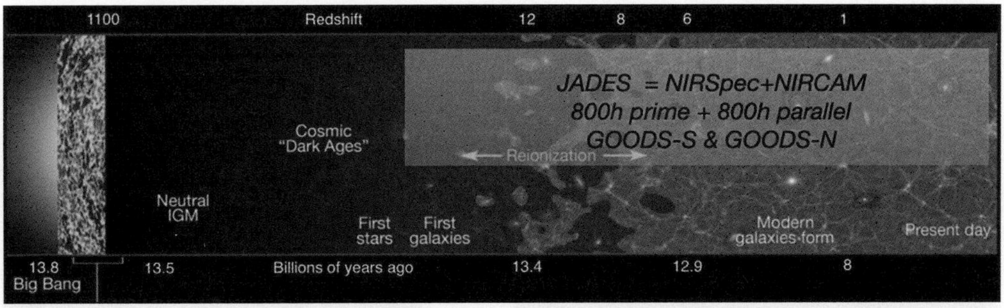

Figure 1. Evolution of the Universe since the Big Bang $\sim$ 13.8 billion years ago through various cosmic epochs, including the recombination era, cosmic dark ages, reionization era, cosmic noon until the present day. The JWST GTO program JADES is designed to obtain images and spectra of galaxies at different redshifts in two deep fields, GOODS-North and GOODS-South, via NIRCam and NIRSpec, respectively. Adapted from Robertson *et al.* (2010).

This work is dedicated to characterizing the populations of LAEs and non-LAEs within the reionization era, by stacking their JWST/NIRSpec spectra based on their properties and redshifts. In the rest of this work, we define LAEs as those Lyα emitting galaxies that have equivalent width (EW) of Lyα emission line > 20Å, while non-LAEs are defined as galaxies with little-to-no Lyα emission. Section 2 presents the data used for this work and the procedure to create stacked spectra. Section 3 presents the main results and conclusions.

2. Data

For this work, we used data from JWST Advanced Deep Extragalactic Survey (JADES; Figure 1 Bunker *et al.* 2020; Rieke *et al.* 2020; Eisentein *et al.* 2023; Bunker et al. 2023). It is the largest JWST Guaranteed Time Observations program combining 800 hours of prime time and 800 hours of parallel time with the Near-infrared Camera (NIRCam) and the Microshutter array (MSA) on NIRSpec, designed to obtain images and spectra of galaxies (Figure 2) in two deep fields initially observed as part of the Great Observatories Origin Deep Survey (GOODS Giavalisco *et al.* 2004), GOODS-North (GOODS-N) and GOODS-South (GOODS-S).

Given that JADES contains deep spectroscopy of various targets at different redshifts, it is suitable for the study of populations of LAEs and non-LAEs. Within JADES, we identified a sample of 42 LAEs with EW(Lya) > 20Å, and a sample of 211 non-LAEs, within a redshift range of 4.8-9.6 such that wavelength covering Lyα and [OIII] lines are included. Figure 3 shows the redshift distribution of all galaxies within the sample (green), LAE sample (grey), and non-LAEs (light blue). We then divide each sample further by the median redshift of LAEs, i.e., 6.3. This gave us a total of six sub-samples of galaxies.

JADES provides the medium-resolution (R1000) spectra of the identified targets within the observed wavelength range of $\sim$0.6–5 microns, taken with prism and medium-resolution gratings on NIRSpec. We use both R100 and R1000 spectra to create the composite spectra. For stacking the R1000 spectra within a sub-sample, we first normalize each R1000 spectrum via the continuum at 1500Å obtained from the corresponding R100 spectrum of that target and then apply a weighted average to all normalized spectra to produce a composite spectrum. For creating a composite spectrum from R100 spectra of a sub-sample, we first normalize each spectrum within the sub-sample by the continua at the [OIII]+Hβ complex. The continuum at the [OIII]+Hβ is too weak to

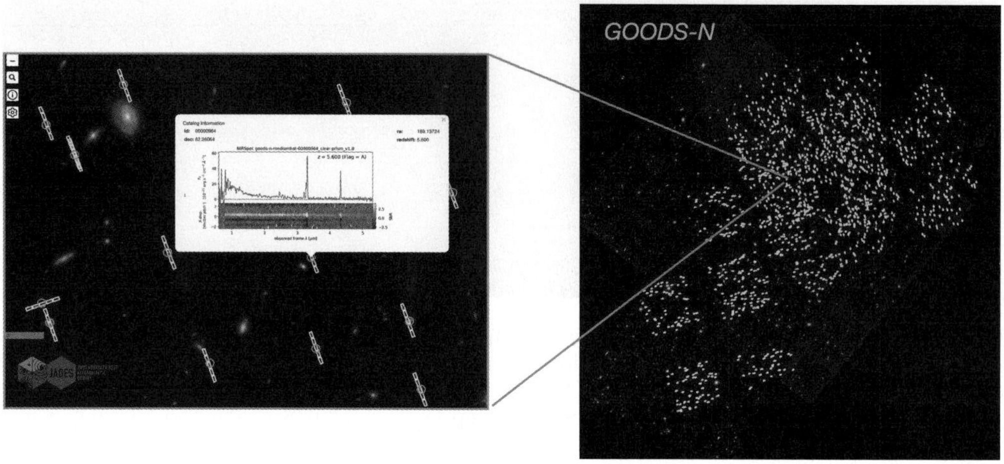

Figure 2. Right-hand panel: A NIRCam color composite image of one of the two deep fields, GOODS-N targeted by the JADES GTO program, where the yellow dots show the slitlet positions created using NIRSpec/MSA shutters to observe galaxies at different redshifts. Left-hand panel: Zoom in on a tiny location in the GOODS-N field, showing the slitlets (yellow) composed of shutters. The inset shows the 1D spectrum (upper panel) and 2D spectrum (lower panel) of a galaxy at redshift z$\sim$5.6 within one of those slitlets and obtained by NIRSpec/MSA Prism. The spectrum shows a plethora of emission lines. The above figure was created by using the interactive viewer located at `https://jades-survey.github.io/viewer/`.

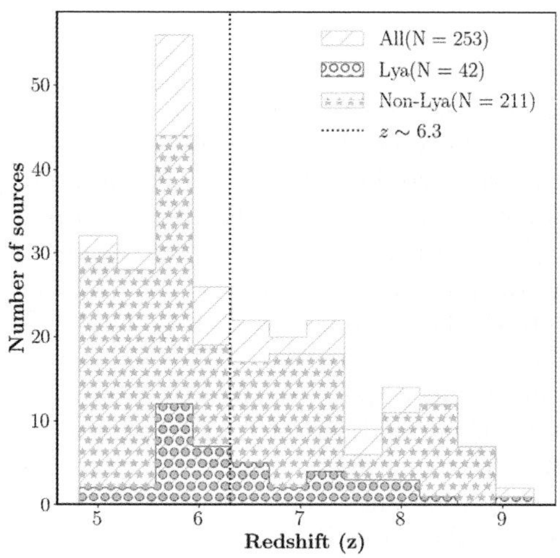

Figure 3. Redshift distribution of all targets (green-hatched histogram), LAEs (grey histogram with circles), and non-LAEs (light blue histogram with stars) used for the analysis in this work. The median redshift of LAEs is $\sim$6.3 (dotted), which is used to further divide the LAEs and non-LAEs populations within two sub-samples. This results in six sub-samples of galaxies based on the EW(Lyα) and redshifts, and are used to create the composite spectra. Reproduced from Kumari *et al.* (2024), with the permission of authors.

be determined in the R100 spectrum of some targets. Hence these targets are excluded. These normalized spectra are then combined by using a weighted average scheme. We thus have six R100 composite spectra and six R1000 composite spectra, corresponding to six sub-samples of LAEs and non-LAEs.

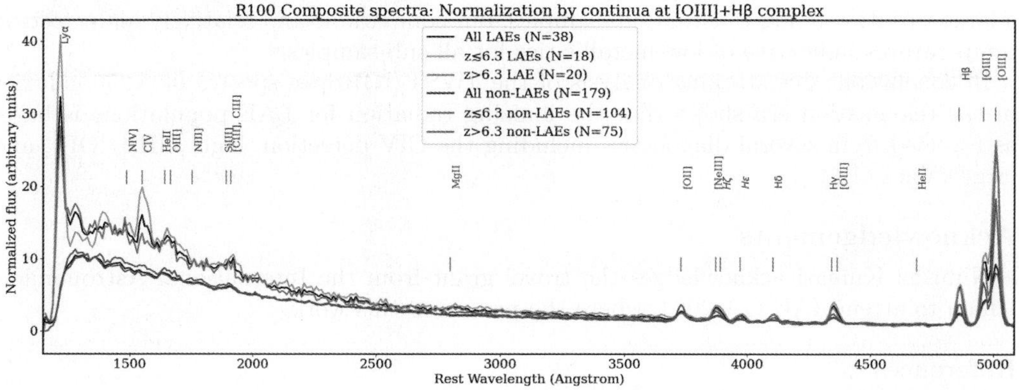

Figure 4. The six R100 composite spectra in rest-wavelength (∼1200-5010Å) corresponding to six sub-samples of LAEs and non-LAEs: (a) all LAEs (black) (b) z$\lesssim$6.3 LAEs (yellow) (c) z>6.3 LAEs (green) (c) all non-LAEs (dark blue) (d) z$\lesssim$6.3 non-LAEs (light blue) (e) z>6.3 non-LAEs (magenta). Within the legend box, N represents the number of sources within each sub-sample. The location of the principal ultraviolet and optical spectral emission lines observed in the galaxies are marked via the brown vertical lines. Reproduced from Kumari *et al.* (2024).

3. Results & Conclusions

Figure 4 shows the six R100 composite spectra produced by following the procedure described in the previous Section 2. We used the R100 composite spectra to estimate the EW([OIII]4959, 5007) for each sub-sample. We find that LAEs show larger EW([OIII]4959, 5007) (>1000Å) compared to non-LAEs irrespective of the redshifts, thus implying a larger ionizing photon production efficiency $\xi^{\star}_{ion}$. Using the EW([OIII]4959, 5007) versus $\xi^{\star}_{ion}$ relation from Chevallard *et al.* (2018), we find that log $\xi^{\star}_{ion}$ ∼ 25.63–25.75 (Hz erg^{-1}) for the three LAEs samples.

The R1000 composite spectra of the six sub-samples of LAEs and non-LAEs show a plethora of UV and optical emission lines, which are resolved due to the relatively higher resolution of medium-resolution gratings of NIRSpec, unlike the case of R100 composite spectra. These resolved lines are useful for studying various physical properties of LAE and non-LAE populations, such as dust extinction, electron temperatures, ionization parameters, and gas-phase metallicities.

We find that the composite spectra corresponding to the LAE samples show CIV 1548, 1550 detection irrespective of the redshifts, whereas CIV is not detected in any non-LAE samples. Detection of high ionization CIV line indicates that LAEs are capable of producing high energy photons.

The UV and optical emission line ratio diagrams are typically used to understand the ionization mechanisms at low redshifts, and the classical line ratio diagrams seem to fail at high redshifts. New line ratio diagnostics are being devised for high-redshifts. We use the diagnostic diagram from Mazzolari *et al.* (2024), including [OIII]4363/Hγ and optical [OIII]5007/[OII]3727, which indicates the AGN might be dominating all samples. Similarly, CIV/Lyα is also indicative of potential AGN activity. We find that CIV/Lyα & CIV/CIII]1907,1909 for LAEs at z$\lesssim$6.3 are ∼ five times that those for LAEs at z>6.3. The LAE populations show much higher [O III]5007/[O II]3728 and low [O II]3728/Hγ at z$\lesssim$6.3 compared to the non-LAE populations, but these line ratios are not sufficient to distinguish the two populations at z>6.3.

We also used [O III]5007/[O II]3728 to estimate ionization parameters (log U) and, [OIII]5007/[OIII]4363 to estimate the electron temperatures for all sub-samples. We find a large variation in log U across sub-samples, but comparable and relatively high electron temperatures indicative of low-metallicities for all sub-samples.

In conclusion, this stacking analysis of the JWST/NIRSpec spectra of >250 galaxies in the reionization era shows that the ionizing radiation for LAE populations is hard, as revealed from several diagnostics including the CIV detection, high [OIII]/[OII], and large values of $\xi^{\star}_{ion}$.

Acknowledgements

Nimisha Kumari acknowledges the travel grant from the International Astronomical Union to attend IAU GA 2024, where she presented this work.

References

Bunker, A. J., *et al.* 2020, *IAUS*, 352, 342
Bunker, A. J., *et al.* 2023, *A&A*, 677, 88
Chevallard, J., *et al.* 2018, *MNRAS*, 479, 3264
Eisenstein, D., *et al.* 2023, *arxiv*, 2306, 2465
Giavalisco, M., *et al.* 2004, *ApJL*, 600, L93
Gareth, G. C., *et al.* 2024, *A&A*, 673, 238
Kumari, N., *et al.* 2024, *arxiv*, 2406, 1199
Mazzolari, G., *et al.* 2024, *arxiv*, 2404, 10811
Partridge, R. B., & Peebles, P. J. E. 1967, *ApJ*, 147, 868
Rieke, M., *et al.* 2020, *IAUS*, 352, 337
Roberts-Borsani, G., *et al.* 2024, *arxiv*, 2403, 7103
Robertson, B. E., *et al.* 2010, *Nature*, 468, 49
Saxena, A., *et al.* 2023, *A&A*, 678, 68
Witstok, J., *et al.* 2024, *arxiv*, 2404, 572

Questions

Themiya Nanayakkara asked to elaborate on the point about the contribution of AGN in the reionization era mentioned in this talk.

The speaker answered that the emission line ratios diagnostic diagram involving [OIII]4363/Hγ and optical [OIII]5007/[OII]3727 (Mazzolari *et al.* 2024), shows that AGN might be dominating all LAE and non-LAE sub-samples presented in this work, though further work needs to be done to understand the role of AGN in creating ionized bubbles in the epoch of reionization.

The First Chapters of Our Cosmic History with JWST
Proceedings IAU Symposium No. 391, 2025
T. Nanayakkara & M. Maseda, eds.
doi:10.1017/S1743921324002667

A merger-driven scenario for clumpy galaxy formation in the epoch of reionization: Physical properties of clumps in the FirstLight simulation

Yurina Nakazato

Department of Physics, The University of Tokyo, 7-3-1 Hongo, Bunkyo, Tokyo 113-0033, Japan
email: `yurina.nakazato@phys.s.u-tokyo.ac.jp`

Abstract. Recent James Webb Space Telescope (JWST) observations with superb angular resolution have revealed the existence of clumpy galaxies at high redshift through the detection of rest-frame optical emission lines. The formation of clumpy galaxies at lower redshift ($z \lesssim 4$) has been investigated both observationally and theoretically, and violent disk instability driven by cold gas accretion is proposed as one of the main mechanisms. However, the formation of clumpy galaxies at $z > 6$ is not well understood theoretically.

In order to study the properties of (sub-)galactic clumps around the epoch of reionization ($z \gtrsim 6$), we use FirstLight, a suite of zoom-in high-resolution ($\sim$20 pc) cosmological simulations. We devise a physical model of HII regions and calculate spatially resolved [OIII] line emission. We identify the giant clumps that are bright in [OIII] 5007Å line with flux greater than $\sim 10^{-18}$ergs^{-1}cm^{-2}, to be detected by JWST. Among 62 simulated galaxies that have stellar masses of $(0.5-6) \times 10^{10}\, M_\odot$ at $z = 5$, we find clumps in 1828 snapshots in the redshift range $z = 5.5 - 9.5$. About one-tenth of the snapshots show the existence of clumpy systems with two or more components. Most of the clumps are formed by mergers and can be characterized by their ages; central clumps dominated by stellar populations older than 50 Myr, and off-centered clumps dominated by younger stellar populations with specific star formation rates of $\sim 50\,\mathrm{Gyr}^{-1}$. The latter type of young clumps is formed from gas debris in the tidal tails of major mergers. The merger-induced clumps are short-lived, and merge within a dynamical time of several tens million years. The number density of the clumpy systems is estimated to be $\sim 10^{-5}$ cMpc^{-3}, which is large enough to be detected in recent JWST surveys.

Keywords. galaxies: high-redshift, galaxies: ISM, galaxies: star formation.

---

1. Introduction

The James Webb Space Telescope (JWST) has significantly advanced our understanding of high-redshift galaxies, particularly during the Epoch of Reionization, i.e., $z = 6 - 9$. With its high sensitivity, JWST has spectroscopically confirmed hundreds of galaxies within these redshifts with strong detection of [OIII] 5007Å (e.g. Roberts-Borsani et al. (2024); Bunker et al. (2024)). The emission features indicate the low-metallicity (0.1–0.2 $Z_\odot$) and highly ionized state ($\log U_{\rm ion} \sim -2$) of the early Universe. The large sample enables us to conduct statistical studies on high-redshift galaxies and provides new insights into their formation and evolution.

Thanks to the high-spatial resolutions of JWST NIRCam/NIRSpec IFU, we can investigate the internal structures of $z \gtrsim 6$ galaxies with a scale of 100 pc. Recent JWST/ NIRCam observations have reported that luminous galaxies at $z \simeq 6 - 8$ show clumpy structures, which had not been resolved by ground-based telescopes or HST (e.g. Hainline et al. (2024); Chen et al. (2023); Harikane et al. (2024); Duan et al. (2024)). Additionally, NIRSpec IFU observations have identified several [OIII] emitters within regions of 10×10 kpc (e.g. Hashimoto et al. (2023)), further suggesting the prevalence of clumpy structures in galaxies at $z > 6$.

Clumpy galaxies have been studied at lower redshifts, where massive clumps on kpc-scales are often observed in disk-like galaxies. Theoretically, there are two dominant scenarios for clump formation: violent disk instabilities (VDI) and mergers. These processes, particularly, VDI and minor mergers, are thought to be the primary mechanisms driving clump formation at $z \lesssim 4$ (e.g. Mandelker et al. (2014, 2017)). As the number of observed clumpy galaxies at $z \gtrsim 6$ increases, it becomes important to investigate the statistical properties of these systems using theoretical simulations. Such analyses allow us to identify the main mechanisms for clump formation in high-redshift galaxies at $z = 6 - 9$.

2. Method

We employed zoom-in cosmological simulations, FirstLight (Ceverino et al. (2017)) to investigate the formation of high-redshift galaxies. The simulations have high-spatial resolutions with maximum resolution of ~ 20 pc and the simulated sample includes all galaxies above a certain mass threshold at $z = 5$ ($M_* \gtrsim 5 \times 10^9\, M_\odot$) within the simulation box of $(40h^{-1}\,\mathrm{cMpc})^3$. Such high resolutions and the completeness of massive galaxies allowed us to conduct not only detailed investigations of individual clumpy structures but also statistical studies of these systems.

In a previous study (Nakazato et al. (2023)), we developed a physical model for HII regions by combining the photoionization code CLOUDY with the computed emission line luminosities. This model enabled us to calculate the [OIII] luminosities for clumps identified in our simulations. To interpret the clumpy morphologies observed by JWST, we focused on large, young clumps based on a threshold surface density of SFR. Specifically, we identified clumps with a minimum radius of ~ 100 pc and an SFR of $\gtrsim 1\, M_\odot \mathrm{yr}^{-1}$. The clumps exhibited [OIII] fluxes on the order of $3 \times 10^{-18}\,\mathrm{erg\,s^{-1}cm^{-2}}$, detectable with reasonable exposure time of 10^4 seconds. The more details are found in Nakazato et al. (2024). By applying this identification method to over 4000 simulation snapshots, we detected a large number of clumps and performed a comprehensive statistical analysis to explore their properties and evolutions.

3. Result

Figure 1 shows examples of clumpy galaxies identified in our simulations. The first column shows the distribution of gas density, and gravitational collapse occurs in high-dense regions. These regions host bursty star formation, which traces the distribution of young stars (2nd column). The clumps exhibit strong [OIII] emission and are detectable by JWST (3rd column). In contrast, the distribution of older stars shows a distinct spatial separation (4th column). For instance, one clump in the lower region of the lower panels does not have a corresponding old stellar component. The dark matter distribution is similar to that of the older stars (5th column). The final columns of Figure 1 show the mass-weighted stellar ages, and we see clear separations between young and old components. Some clumps are younger than 30 Myrs, while others have ages up to 300 Myrs, suggesting the existence of distinct clump populations (6th column).

We analyze the physical properties of clumps and categorize them into two categories based on their age. Clumps younger than 50 Myrs are typically gas-rich and exhibit high specific star formation rates (sSFR), located in environments with a high baryonic

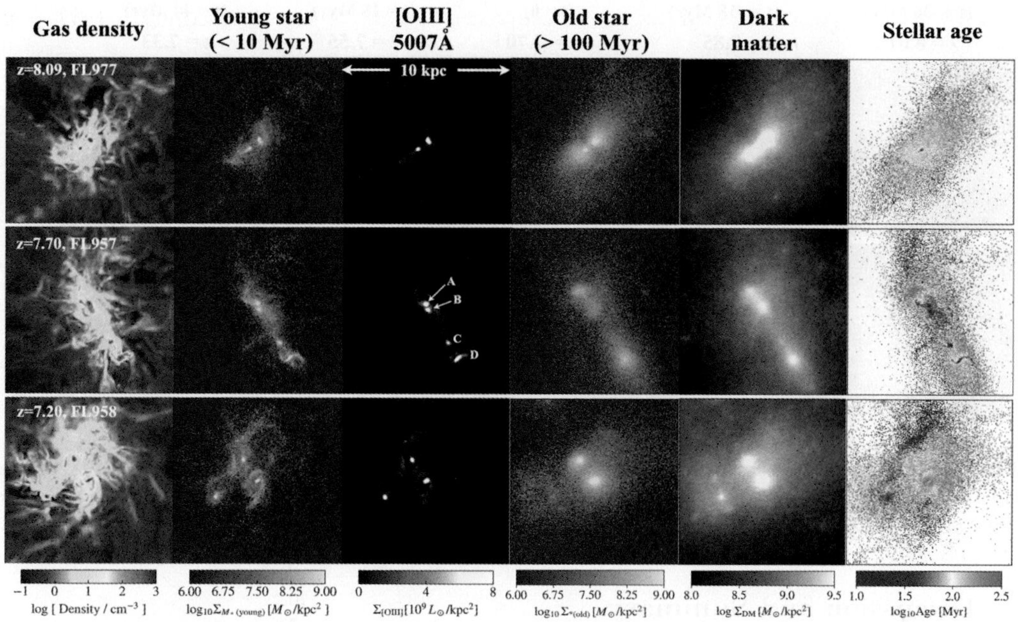

Figure 1. Three simulated clumpy galaxies are shown at redshifts $z = 9.00$ (top, ID = FL977), $z = 7.70$ (middle, ID = FL957), and $z = 7.20$ (bottom, ID = FL958). The columns, from left to right, represent the average gas density, surface stellar mass density of the young population (< 10 Myr), [OIII] 5007 Å line flux, surface stellar mass density of the old population (> 100 Myr), dark matter distribution, and mass-weighted stellar age. Each panel covers a 10 proper kpc square region with the same projected depth across all columns.

fraction. In contrast, older clumps (those older than 50 Myrs) have different characteristics, with less gas and lower sSFRs, indicating different evolution paths between the two populations.

We follow the time evolutions of clumpy galaxies at $z > 6$, and find that the clumps are formed primarily by major mergers. Young clumps in clumpy galaxies form through gas debris in tidal tails produced during the mergers. Figure 2 represents the formation and evolution of clumpy systems for one galaxy sample (middle-row panels in Figure 1). The main galaxy is located at the center, while a secondary dark matter halo enters the region, leading to clump formation in the tidal tails following the peri-center passage. Our analytical estimation suggests that the stellar mass ratios required to form these observable components correspond to major mergers. These findings indicate that the giant, young clumps observed at high redshift are predominantly the result of mergers. Regarding the fate of these clumps, our simulations show that they coalesce into the central proto-bulge within a few tens of Myrs. Initially, after the peri-center passage, multiple clumps form. About 20 Myrs after the formation of four clumps, these clumps merge into two larger structures. Further 27 Myrs later, they coalesce into a single massive system. This process indicates that the clumpy phase is temporary, and clumps eventually merge to form more stable, larger systems within a dynamical time.

Finally, we calculate the number density and clumpy fraction of these galaxies. We find that approximately 10-20% of galaxies with stellar masses greater than $10^9\,M_\odot$ exhibit clumpy morphologies. The number density of clumpy structures is estimated to be $\sim 10^{-5}$cMpc^{-3}, consistent with the major merger rate density at $z = 6 - 9$ (Rodriguez-Gomez et al. (2015)).

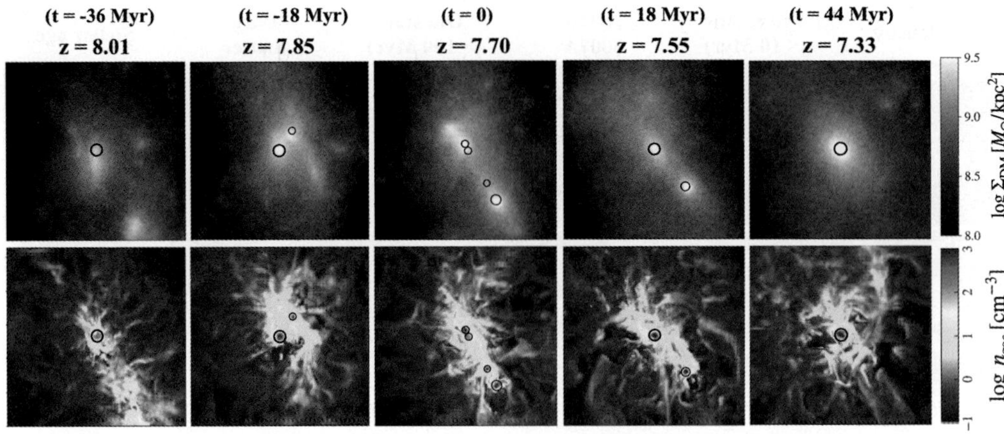

Figure 2. Time evolution of FL975 (middle row in Figure 1). The top and bottom panels show the dark matter density and gas density distribution from $z = 8.01$ to $z = 7.33$. Both projections have side and projected lengths of 10 kpc. Black circles indicate the positions of the identified clumps. Time $t = 0$ is set at $z = 7.70$, when the galaxy consists of four luminous clumps.

4. Discussion and Summary

We have studied the formation of clumpy galaxies in the epoch of reionization observable in rest-frame optical emission lines with a reasonable exposure time. The clumps we identify are bright in [OIII] 5007 Å, with radii of $\sim$ 150-200 pc, which can be resolved by JWST NIRCam/NIRSpec IFU with a reasonable exposure time. Our results show that large clumps form mainly through major mergers. These clumps are categorized into two types: proto-buldges with stellar populations older than 50 Myr, and off-centered clumps dominated by younger stars. The latter ones are formed from gas debris in tidal tails during mergers. The merger-driven clumps have high gas fractions ($> 90\%$) and trigger bursty star formation with sSFR of $\sim 50\,{\rm Gyr}^{-1}$. About 10-20 % of galaxies with $M_* \geq 10^9\,M_\odot$ at $z = 6 - 9$ are clumpy, and these systems are short-lived, merging within tens of Myrs. The number density of clumpy systems is $\sim 10^{-5}\,{\rm cMpc}^{-3}$, large enough to be observable in the JWST survey area.

We find that clumps at $z = 6 - 9$ are predominantly merger-driven, while clumps at lower redshifts ($z \lesssim 4$) often form through VDI, as studied in previous works (e.g. Mandelker et al. (2017)). Interestingly, VDI is less likely to occur in low-mass galaxies due to their small disk masses (Dekel et al. (2009)). Even when VDI does occur, the resulting clumps typically hold only 1-2% of the central object's mass (Ceverino et al. (2010)), making them too faint to be detected by current observations. Our findings may imply a shift in clump formation mechanisms, transitioning from a predominantly merger-driven process in the early universe to VDI becoming more prominent at lower redshifts ($z \lesssim 4 - 5$).

Some limitations still remain in our study. Firstly, our study uses a fixed surface star formation rate threshold, which might miss small clumps in low-mass galaxies. Some of these unidentified clumps could be proto-globular clusters, as suggested by recent JWST observations of young stellar clusters at $z = 6 - 10$ (see section 5).

Secondly, our clump identification is based on optical emission lines, but multiwavelength studies (UV to NIR) reveal spatial offsets between UV, optical, and FIR emissions in clumpy galaxies. Clumpy galaxies formed by major mergers start burst star formation and simultaneously produce large amounts of dusts. To consider dust attenuation, further post-processing Monte Carlo dust radiative transfer is performed on each snapshot of the simulated galaxies. Our analysis shows that dust attenuation has a small impact on the rest-frame optical emissions, which effectively trace the dense-gas clumps formed by mergers. However, rest-frame UV emissions are significantly affected by inhomogeneous dust attenuation. We find that UV luminosities from gas clumps are strongly dust-attenuated due to the large column density. The surrounding more diffuse gases have low intrinsic UV luminosity but less attenuation due to their small column density, leading to be observed as bright components. This results in rest-frame UV images appearing more clumpy compared to their rest-frame optical counterparts, and UV-identified clumps do not necessarily trace gravitationally bound objects. We will further discuss this in a future paper (Nakazato et al. in prep.).

Our findings in this study provide new insights into the formation and evolution of high-redshift galaxies, helping us understand the complex processes in the early universe. The combination of high-resolution simulations and observational data from JWST reveals more detailed physical properties of clumpy galaxies in multiwavelengths.

5. Q & A

In this section, we attach questions and answers that followed our talk at the IAU Symposium 391.

Q: *Can physical processes such as stellar winds, cosmic rays, or AGN feedback play a significant role in suppressing the formation of clumps in high-redshift galaxies?*
A: The FirstLight simulations we used in this study do not include AGN feedback or cosmic rays, thus we cannot assess the effect quantitatively. However, we roughly estimate these effects in the following discussions.
For AGN feedback, the timescale is assumed to be roughly the same as BH growth timescales, i.e., Salpeter timescale

$$t_{\rm Sal} = \frac{M_{\rm BH}}{\dot{M}_{\rm Edd}} = \frac{\sigma_{\rm T} c \epsilon_{\rm R}}{4\pi G m_{\rm p}} = 45\,{\rm Myr}\left(\frac{\epsilon_{\rm R}}{0.1}\right), \tag{1}$$

where $\sigma_{\rm T}, \epsilon_{\rm R}$ are the cross section of Thomson scattering and the radiative efficiency. The average clump gas density in our simulation is $\sim 250\,{\rm cm}^{-3}$, and the corresponding free-fall time scale is

$$t_{\rm ff} = \sqrt{\frac{3\pi}{32}}\frac{1}{\sqrt{G\rho}} \sim 3.2\,{\rm Myr}\left(\frac{n_{\rm gas}}{250\,{\rm cm}^{-3}}\right)^{-1/2}. \tag{2}$$

Therefore, the clump formation (star formation in clumps) can occur before the AGN feedback becomes dominant.

For stellar winds, we consider a radiation pressure from newly formed stars in HII regions. To assess such feedback effect on clump survival qualitatively, we follow the calculation of Krumholz and Dekel (2010) and Fall et al. (2010);

$$\varepsilon = \frac{M_*}{M} = \frac{\Sigma}{\Sigma + \Sigma_{\rm crit}}, \tag{3}$$

$$\Sigma_{\rm crit} = \frac{5\eta f_{\rm trap}\langle L/M_*\rangle}{\pi \alpha_{\rm crit} G c}. \tag{4}$$

Here ϵ is the final stellar fraction, and both N-body simulations and analytic models argue that if $\epsilon \gtrsim \epsilon_{\rm th} = 0.5$, then most of the stellar mass remains bound, not be destructed by radiation feedback. Also, η is a constant that depends on the density distribution within the clump, $f_{\rm trap}$ is the factor that represents the radiation force is enhanced by the reradiated IR within the expanding shell, and $\alpha_{\rm crit}$ is a parameter for the critical velocity to eject mass from the cloud. We adopt the fiducial values of $\eta = 2/3$ and $f_{\rm trap} = \alpha_{\rm crit} = 2$, which represents local star forming regions (Fall et al. (2010)). For the light-to-stellar mass ratio, $\langle L/M_*\rangle$, we adopt $\langle L/M_*\rangle = 377\,{\rm erg s^{-1} g^{-1}}$ for 10 Myrs of bursty star formation calculated using BPASS SED (Eldridge et al. (2017)). Substituting these values into Equation (4), we obtain $\Sigma_{\rm crit} = 0.2\,{\rm gcm}^{-2}$. The surface mass density of gas clumps is calculated by using the average gas density ($250\,{\rm cm}^{-3}$) and the average radius (140 pc) of newly formed clumps in our simulations, and derived as $\Sigma = 0.24\,{\rm gcm}^{-2}$. By substituting these values into Equation (3), we obtain $\epsilon = 0.55$, which exceeds the critical value ($\epsilon_{\rm th}$). Then the formed clumps are expected to remain. We notice that our analysis focuses on giant clumps with radii of $\sim$ 100 pc. Stronger feedback models can prevent the formation and survival of smaller clumps or stellar clusters.

For cosmic rays (CR) feedback, CR loses the energy by Coulomb and ionization interactions and heats the thermal gas (Enßlin et al. (2007)), which leads to increasing of effective pressure of gas and suppressing star formation. Jubelgas et al. (2008) perform cosmological simulations including CR physics, and calculate the contribution of the CR pressure to the total pressure ($P_{\rm CR}/(P_{\rm CR} + P_{\rm thermal})$). We compare their results at $z = 6$ and find that the CR pressure contribution can be expected to be less than 10% for our clump gas density($\sim 250\,{\rm cm}^{-3}$), which implies CR does not affect the star formation in the formed clumps.

Q: *In your clump formation scenarios, are clumps always formed within tidal tails?*
A. We have checked all clumpy galaxies with multiple components (i.e., $N_{\rm clump} \geq 3$) at $z = 6-9$, and confirmed that the third/fourth components are formed by tidal debris through major mergers. Some clumpy galaxies at $z < 6$ formed multiple *observable* clumps by violent disk instabilities but these snapshots are out of range of our analysis.

Q. *How do the physical scales of your simulated clumps compare to recent strong lensing observations from JWST (e.g., the $z = 8.3$ clumpy galaxy reported by Mowla et al. 2024)?*
A. Since our simulations have the minimum cell size of 17 pc (Ceverino et al. (2017)) and set additional pressure floors to prevent artificial fragmentation, we can only follow the growth of structures above $\sim$ 100 pc scales. Some cosmological simulations, such as those by Ma et al. (2018); Calura et al. (2022); Garcia et al. (2023), focus on the formation of dwarf galaxies and stellar clusters at $z > 6$, achieving sub-pc scale resolution. This high resolution enables the investigation of lensed stellar clusters, which are recently observed by JWST (Mowla et al. (2024); Adamo et al. (2024); Bradley et al. (2024)) and have effective radii of less than 1 pc.

Q: *Were the clumps produced in your simulations short-lived? If they were created by mergers, could their presence be used to estimate the timing of these mergers?*
A: Yes, the clumps in our simulations are short-lived, typically merging within a few tens of Myr, approximately on the order of the system's dynamical time. During a major merger, clumpy morphologies are identified at various stages, such as during the initial encounter, peri-center passage, and multiple clump formation, with the entire process spanning roughly 80 Myr (also see Figure 2). If multiple clumps are observed, it likely indicates that a halo encounter occurred about 30–40 Myr ago with a separation of $\lesssim$ 5 kpc.

References

Roberts-Borsani, G., et al. 2024, arXiv: 2403.07103
Bunker A. J., et al., 2023, *A&A*, 677, A88
Hainline, K. N., et al., 2024, *ApJL*, 964, 1
Chen, Z., et al.,2023, *MNRAS*, 518, 5607
Harikane, Y., et al., 2024, arXiv:2406.18352
Duan, Q., et al., 2024, arXiv:2407.09472
Hashimoto, T., et al., 2023, *ApJL*, 955, L2
Mandelker, N., et al., 2014, *MNRAS*, 443, 3675

Mandelker, N., et al., 2017, *MNRAS*, 464, 635
Ceverino, D., Glover, S. C. O., & Klessen, R. S. 2017, *MNRAS*, 470, 2791
Nakazato, Y., Yoshida, N., & Ceverino, D. 2023, *ApJ*, 953, 140
Nakazato, Y., Ceverino, D., Yoshida, N., 2024, arXiv:2402.08911
Adamo, A., et al., 2024, *Nature*, 632, 8025
Mowla, L., et al. 2024, arXiv:2402.08696
Ma, X., et al. 2018, *MNRAS*, 477, 219
Calura, F., et al. 2022, *MNRAS*, 516, 5914
Garcia, F. A. B., Ricotti, M., Sugimura, K., & Park, J., 2023, *MNRAS*, 522, 2495
Bradley, L. D., et al. 2024, arXiv:2404.10770
Dekel, A., Sari, R., & Ceverino, D. 2009, *ApJ*, 703, 785
Ceverino, D., Dekel, A., & Bournaud, F. 2010, *MNRAS*, 404, 2151
Fall, S. M., Krumholz, M. R, Matzner. C. D, 2010, *ApJL*, 710, 2
Krumholz, M. R, Dekel, A., 2010, *MNRAS*, 406, 1
Jubelgas, M., Springel, V., Enßlin, & Pfrommer, C., 2008, *A&A*, 481, 1
Enßlin, T. A., Pfrommer, C., Springel, V., Jubelgas, M., 2007, *A&A*, 473, 1
Eldridge, J. J., Stanway, E. R., Xiao, L., et al. 2017, *PASA*, 34, e058
Rodriguez-Gomez, V., Genel, S., Vogelsberger, M., et al., 2015, *MNRAS*, 449, 49

The First Chapters of Our Cosmic History with JWST
Proceedings IAU Symposium No. 391, 2025
T. Nanayakkara & M. Maseda, eds.
doi:10.1017/S1743921324002278

The youngest sibling of the Milky Way: JWST reveals a barred spiral galaxy at $z \sim 3$

Luca Costantin

Centro de Astrobiología, CSIC-INTA, Ctra. de Ajalvir km 4, Torrejón de Ardoz, E-28850, Madrid, Spain
email: lcostantin@cab.inta-csic.es

Abstract. Understanding the cosmic evolution of the barred galaxy population is fundamental to quantifying their impact on the internal secular evolution of galaxies and dating the formation of the first dynamically cold stellar disks in the Universe. In this context, the new imaging capability provided by JWST is revolutionizing the study of high-z disk galaxies, unveiling for the first time their complex morphologies. In this contribution, I discuss the discovery of ceers-2112, a Milky Way-like barred spiral galaxy observed in the CEERS survey. The morphological analysis of the stacked NIRCam images undoubtedly discloses the presence of a stellar bar and hints of spiral arms. The combined analysis of HST/ACS, HST/WFC3, and JWST/NIRCam photometry places the galaxy at redshift $z = 3.03$. Thus, this bar was already formed when the Universe was only 2 Gyr old. The star formation history of ceers-2112 reveals that this system is a Milky Way-like galaxy, with $M_\star = 4 \times 10^9\ M_\odot$. Remarkably, this discovery proves that mature stellar (and not only gaseous) disks were in place as early as ~ 12 Gyrs ago ($z \sim 5$) and that baryons were dominating over dark matter in ceers-2112 at these early times.

Keywords. Galaxies formation, Galaxies evolution, Galaxies morphology

1. Introduction

The first morphological studies at Cosmic Noon carried out with the James Webb Space Telescope (JWST) are remarkable in several ways. The rest-frame optical structure of galaxies at $z > 2$ suggests that the morphological diversity seen in the local Universe is already in place as early as $z \sim 5-6$ (Ferreira *et al.* 2022; Kartaltepe *et al.* 2023; Costantin *et al.* 2024). Contrary to previous expectations, disk galaxies represent the dominant morphological type at these redshifts, even if the intrinsic nature of these systems is still debated (Vega-Ferrero *et al.* 2024; Pandya *et al.* 2024). At $2 < z < 4$, an unexpected fraction (up to $\sim 15\%$) of disk galaxies with $M_\star > 10^{10} M_\odot$ hosts well-developed stellar bars similar to their mature low-z counterparts (Guo *et al.* 2023; Le Conte *et al.* 2023; Guo *et al.* 2024). In this context, these first barred galaxies formed in the Universe are fundamental laboratories for unveiling the interplay between baryonic and dark matter at early times. Furthermore, their discovery is key to identify the epoch when the first dynamically cold stellar disks formed in the history of the Universe (Bland-Hawthorn *et al.* 2023). Indeed, the search and characterization of barred galaxies in the early Universe provide a complementary route to confirm previous ALMA results targeting only the gaseous component of high-z galaxies and discovering rotationally-supported disks at $z > 4$ and as early as $z \sim 7$ (Rizzo *et al.* 2020; Rowland *et al.* 2024).

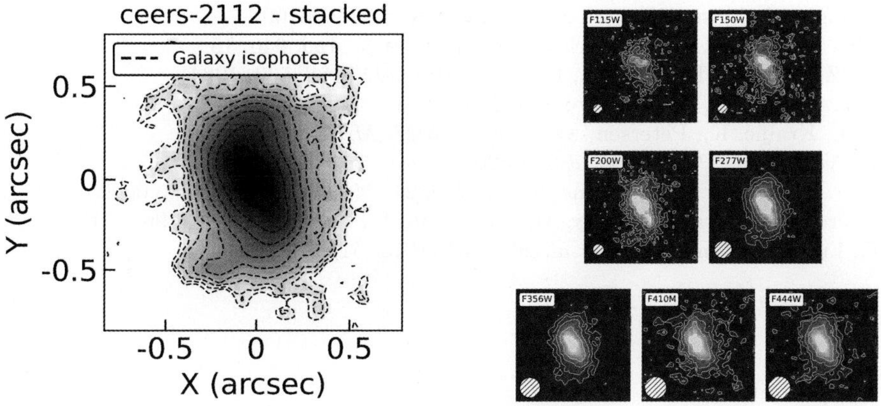

Figure 1. ceers-2112 stacked image (left panel) and postage stamps in 7 NIRCam bands (right panels). Adapted from Costantin *et al.* (2023).

2. Summary of the main results

Among the population of high-z disk galaxies, there is ceers-2112 (Figure 1), the only Milky Way-like barred spiral galaxy observed to date beyond $z = 2$ (Costantin *et al.* 2023). The barred nature of the galaxy was identified thanks to imaging observations taken during the first epoch (21-22 June 2022) of the CEERS campaign (Finkelstein *et al.* 2022; Bagley *et al.* 2023). The morphological modeling of the galaxy reveals the presence of a bar component with strength $S_{\rm bar} = 0.23 \pm 0.01$ and length $r_{\rm Ferrers} = 0.42 \pm 0.03$ arcsec (3.3 kpc). The HST/ACS, HST/ WFC3, and JWST/NIRCam photometry is used to infer that ceers-2112 has a photometric redshift of $z = 3.03^{+0.04}_{-0.05}$. The star formation history of ceers-2112 tells us that it can be considered a progenitor of the Milky Way, having a stellar mass of $M_\star = 4 \times 10^9\ M_\odot$ and mass-weighted age of 620^{+150}_{-160} Myr. This barred galaxy was already mature when the Universe was only 2 Gyr old, challenging cosmological simulations that predict stellar bars to be almost absent at $z > 1.5$ in the progenitors of Milky Way-like galaxies (Zana *et al.* 2022; Reddish *et al.* 2022). Considering that stellar bars form in timescales greater than 1 Gyr (Fujii *et al.* 2018), these findings also pose severe challenges to our current understanding of bar formation.

3. Conclusions

The discovery of ceers-2112 suggests that dynamically cold stellar disks are in place as early as $z = 5$ and that stellar bars could form in less than 500 Myr, challenging the current understanding of the formation timescale of bar structures. Finally, this discovery provides strong observational constraints about baryons and dark matter interplay at $z > 3$, proving that baryon-dominated disks should be in place at these early times, with a fraction of stars and gas over dark matter greater than 60–70%.

References

Bagley, M. B., Finkelstein, S. L., Koekemoer, A. M., *et al.* 2023, *ApJ*, 946, 12
Bland-Hawthorn, J., Tepper-Garcia, T., Agertz, O., *et al.* 2023, *ApJ*, 947, 80
Costantin, L., Pérez-González, P.-G., Guo, Y., *et al.* 2023, *Nature*, 623, 499
Costantin, L., Gillman, S., Boogaard, L. A., *et al.* 2024, *A&A*, arXiv:2407.00153
Ferreira, L., Nathan, A., Conselice, C. J., *et al.* 2022, *ApJ*, 938, 2
Finkelstein, S. L., Bagley, M. B., Arrabal Haro, P., *et al.* 2022, *ApJ*, 940, 55
Fuji, M. S., Bédorf, J., Baba, J., *et al.* 2018, *MNRAS*, 477, 1451
Guo, Y., Jogee, S., Finkelstein, S. L., *et al.* 2023, *ApJ*, 945, 10

Guo, Y., Jogee, S., Wise, E., *et al.* 2024, *ApJ*, arXiv:2409.06100
Kartaltepe, J. S., Rose, C., Vanderhoof, B., *et al.* 2023, *ApJ*, 946, 15
Le Conte, Z., Gadotti, D., Ferreira, L., *et al.* 2024, *MNRAS*, 530, 1984
Pandya, V., Zhang, H., Huertas-Company, M., *et al.* 2024, *ApJ*, 963, 54
Reddish, J., Kraljic, K., Petersen, M. S., *et al.* 2022, *MNRAS*, 512, 160
Rizzo, F., Vegetti, S., Powell, D., *et al.* 2020, *Nature*, 584, 201
Rowland, L. E., Hodge, J., Bouwens, R., *et al.* 2024, *MNRAS*, arXiv:2405.06025
Vega-Ferrero, J., Huertas-Company, M., Costantin, L., *et al.* 2024, *ApJ*, 961, 51
Zana, T., Lupi, A., Bonetti, M., *et al.* 2022, *MNRAS*, 515, 1524

The First Chapters of Our Cosmic History with JWST
Proceedings IAU Symposium No. 391, 2025
T. Nanayakkara & M. Maseda, eds.
doi:10.1017/S1743921324002394

Star formation in Illustris TNG galaxy mergers

András Péter Joó[1], Bendegúz Koncz[1,3], Andreas Burkert[2], István Horváth[4] and L. Viktor Tóth[1,3]

[1]Institute of Physics and Astronomy, Eötvös Loránd University, Budapest, Hungary

[2]Universitäs-Sternwarte, Ludwig-Maximilians-Universität München, Germany

[3]Faculty of Science and Technology, University of Debrecen, Hungary

[4]University of Public Service, Budapest, Hungary
email: `andraspeterjooastro@gmail.com`

Abstract. Cosmological magneto-hydrodynamical simulations such as IllustrisTNG describe the first chapters of our cosmic history. In our study, we focus on the evolution of the physical and chemical properties of galaxies in a series of mergers. We processed the catalogue data and the merger tree files of the TNG100-1 and the TNG300-1 simulations and investigated the evolution of progenitor galaxies, their merger companions (next progenitor) and their common descendants. We calculated the galaxies' average star formation rate (SFR), metallicity and mass at different redshifts between $0 < z < 15$ and found that the SFR of the descendant galaxies is increasing through the mergers. We traced both the large-scale structure and mass build-up in events involving dwarf galaxies. We reconstructed the observed cosmic star formation rate density variation through a large time interval ($12 > z > 0$) and confronted the model predictions with recent JWST observations. We discuss how the dwarf galaxies influenced the mass increase of galaxies from both observational and computational approaches. Our results show that a high fraction of galaxy mergers is connected to superclusters, especially at lower redshifts.

Keywords. star formation rate density, cosmic star formation history, galaxy evolution, JWST

1. Introduction

Star formation is a fundamental process that shapes galaxies and, consequently, the Universe itself. Understanding how star formation in galaxies changes across cosmic time is essential for constructing a comprehensive view of galaxy evolution. Looking back in time at cosmic scales, the star formation rate (SFR) of galaxies changes as illustrated e.g. in the well-known Madau diagram of cosmic star formation history (CSFH, Madau *et al.* (2014)). The change in a galaxy's SFR results from a combination of internal and external processes, with galaxy mergers being one of the key external factors.

Mergers play a key role in the lifecycle of galaxies, contributing significantly to their mass build-up and potentially influencing star formation.

Suess *et al.* (2023) studied the mass buildup in galaxy mergers using the JWST Advanced Deep Extragalactic Survey, revealing a large population of low-mass companions around massive quiescent galaxies. They observed galaxies with stellar masses of $\log(M_*/M_\odot) > 10.5$ at redshift between 0.5 and 3, identifying 161 massive galaxies with 629 companions of which only 10% have a mass of $\log(M_*/M_\odot) \geq 9$ and only 6% have mass ratios >1:10. They found that the average massive quiescent galaxy hosts approximately five nearby companions with stellar mass ratios <1:10, and despite a median stellar mass ratio of 1:900, these tiny companions account for at least 30% of the total added mass.

The effect of mergers on star formation is a widely studied topic, yet the nature of this impact remains debated. Some mergers trigger star formation activity (e.g. by Hwang *et al.* (2019) and Mihos *et al.* (1996)), other mergers show only a minor effect on star formation rates (e.g. Pearson *et al.* (2019)), and in many cases, the interaction can lead to quenching (e.g. Moustakas *et al.* (2013), Pontzen *et al.* (2017)). An open question remains as to how mergers impact the average star formation rate across the galaxy population. One effective method for investigating this question is the use of cosmological simulations.

IllustrisTNG (Pillepich *et al.* (2018), Marinacci *et al.* (2018), Naiman *et al.* (2018), Nelson *et al.* (2018), Springel *et al.* (2018)) is a suite of large-volume, cosmological, gravo-magneto-hydrodynamic simulations including a comprehensive model for galaxy formation. Each TNG simulation self-consistently solves for the coupled evolution of dark matter, cosmic gas, luminous stars, and supermassive black holes from redshift z = 127 to 0 and generates 100 resulting snapshots from z = 20 to 0 with approximately uniform time resolution (150$\pm$50 million years). Galaxies are identified as subhalos of particles associated with each other in each snapshot, and the calculated properties of these subhalos are stored in catalogue files. The merger tree catalogue files contain the relations of merging galaxies (progenitors and descendants) besides all of the calculated galaxy properties. Each simulation run got several independent merger trees stored in separate files. As we concluded in one of our previous studies, these files are independent, and the galaxy properties don't differ significantly in different merger trees (Koncz *et al.* 2023). Two of the simulation runs we also used for analysis are TNG100-1 and TNG300-1. The middle-sized TNG100-1 has a volume of 110.7^3 Mpc^3 and contains $\approx$4.4 Million galaxies at z = 0, while the largest-sized run that is currently available, TNG300, has a volume of 302.6^3 Mpc^3 containing $\approx$14.5 Million galaxies at z = 0. The merger trees of these runs make it possible to analyse star formation rates of merging galaxies through cosmic time scales.

The results of simulations can provide predictions based on current models. To assess the validity of these models, we can compare them with observational data. Over cosmic timescales, this comparison is only feasible with the inclusion of high-redshift observations. The James Webb Space Telescope (JWST, Gardner *et al.* (2006), Kalirai *et al.* (2018)) enabled the study of galaxies farther away than ever before. Recent papers examined the mass build-up of galaxies up to z = 3 (Suess *et al.* 2023) and the SFR history of galaxies as far as up to z = 8 (Kim *et al.* 2023). These results provide an opportunity to test the predictions of simulations over the longest timespan to date.

We studied galaxy mergers in the IllustrisTNG cosmological simulation, tracing the star formation rate (SFR) and mass of the merging galaxies over the simulation's time span. We examined the effect of mergers on star formation and mass build-up. We also constructed the simulation's cosmic star formation history and compared it to recent JWST results.

We are currently preparing a paper on our extended analysis of star formation in IllustrisTNG galaxy mergers (Koncz *et al.* in. prep.). Here, we highlight key findings in relation to recent JWST results.

2. Data analysis

We selected the TNG100-1 run for constructing the SFR history of merger galaxies, as this made it feasible for us to gather SFR and mass data for all galaxies across each of the 100 snapshots from z = 20 to 0, achieving the highest time resolution in the process. We calculated averages for all galaxies within each snapshot.

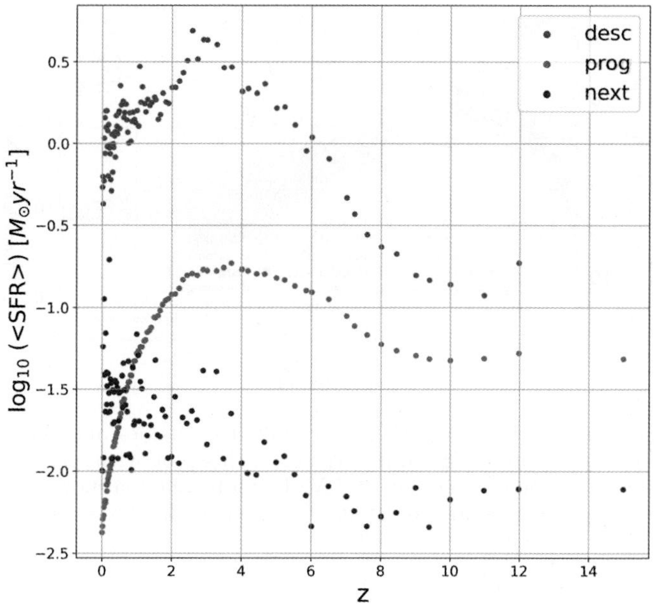

Figure 1. Average SFR of merger galaxies versus redshift z. First progenitor galaxies (first - the most massive ones) are shown in red, next progenitors (next) in blue, and their descendants (desc) in green. SFR of the descendant galaxies have higher values at all redshifts than the progenitors, showing the effect of the mergers on average. The peak of the curve is between z = 2 and 3, where it is two orders of magnitude higher than the first progenitors' value.

To analyse CSFH and mass build-up, we selected TNG300-1 to have the largest volume available, in which we only processed the snapshots with integer-valued redshifts (or very close to them) to save runtime, as high time resolution was not needed for this analysis.

For both volumes, we processed one of the merger trees to save computing resources. We selected three types of galaxies taking part in a merger event based on their relation: first progenitor (first) and next progenitor (next), the largest and second-largest galaxies by mass joining in a merger, and descendant (desc), the resulting galaxy. We calculated averages for these types separately.

For studying mass build-up by mergers we selected a sample of large quiescent galaxies from the TNG300-1 run's first merger tree at redshift z = 0. For selection criteria we used $\log(\mathrm{sSFR}) < -10.5$ and $\log(\mathrm{M}_*/\mathrm{M}_\odot) > 10.6$. We gathered the properties of all infalling galaxies for all snapshots, excluding snapshots above z = 8 as these contained only small amounts of mergers or none. In this paper, we highlight one of the selected quiescent galaxies (ID 1094694), which is average-sized in the sample with a mass of $2.7 \cdot 10^{12}\mathrm{M}_\odot$ at z = 0.

To compare the star formation history in TNG with JWST results, we calculated the star formation rate density (SFRD) using the volume of TNG300-1. We used the same snapshots as a basis for the SFRD analysis as we did for the mass build-up analysis, while also adding a few data points around the peak to gain a more precise view of its location. We calculated the average SFR for all galaxies in these snapshots, not just those involved in mergers.

3. Results and discussion

Figure 1 shows the SFR history of TNG merger galaxies. For the larger galaxies (first progenitors and descendants) it resembles the trend of the cosmic star formation history

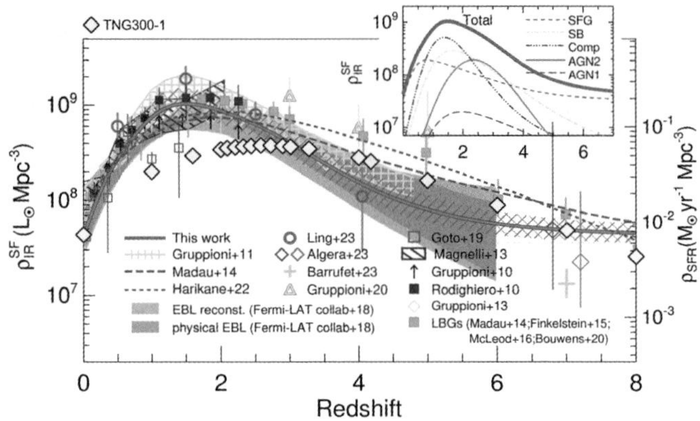

Figure 2. Comparison between the SFRD of TNG300-1 galaxies and the CSFH results from Kim *et al.* (2023) derived from JWST data. We used this plot with the first author's consent and overplotted our data points. The recent JWST results show a more prominent peak ≈0.5 dex higher, and also earlier, around a redshift of 1.5, while the TNG model shows a shallower curve with a maximum between z = 2 and 3.

(CSFH) according to Madau *et al.* (2014), however, peaking a bit later between redshifts z = 2 and 3. In contrast, the SFR history of the next progenitors shows a rather continuously rising trend, however with big fluctuations. We attribute the fluctuations of data points to the relatively small volume and sample size. The average SFR of descendants is approximately 1 dex higher than that of the first progenitors at redshifts above 6-7, with this difference increasing to 1.5-2 dex at lower redshifts, indicating a clear effect of mergers elevating SFR on average. Notably, below z≈1, the average SFR of the next progenitors surpasses that of the first progenitors. This may be due to the concentration of most galaxy mass within a few massive galaxy clusters, primarily composed of elliptical galaxies with low SFRs, into which merging galaxies fall. Further analysis, however, is required to clarify this effect.

Figure 2 compares the SFRD history of TNG300 galaxies with the recent CSFH results from Kim *et al.* (2023) derived from JWST data. The JWST results show a more pronounced peak in SFR density, approximately 0.5 dex higher and appearing earlier, around a redshift of 1.5. In contrast, the TNG model produces a shallower curve, aligning more closely with the previously established model of Madau et al. (2014), though with a peak between redshifts 2 and 3. We explore these differences further in our forthcoming paper.

Figure 3 presents the distribution of merger mass ratios for the selected large, quiescent galaxy (ID 1094694), covering all merger events it has experienced. All mergers occurred with mass ratios below 1:10, with most even below 1:1000. We found similar results across our sample galaxies, aligning with Suess *et al.* (2023), who reported a median stellar mass ratio of approximately 1:900 for merger companions.

Figure 4 shows the mass distribution of mergers for our selected large quiescent galaxy (ID 1094694). A significant fraction of infalling galaxies have a mass $\log(M_*/M_\odot) \geq 9$, in contrast to Suess *et al.* (2023), who found only 10% in their field. Looking at other galaxies in our sample, a fraction of 30–40% of the merger companions have a mass $\log(M_*/M_\odot) \geq 9$. We will investigate this difference in depth in our forthcoming paper too.

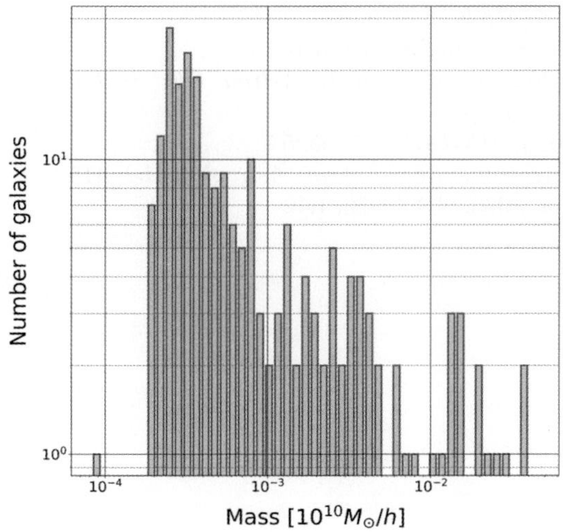

Figure 3. Mass ratio distribution of the merger companions of our selected large, quiescent galaxy (ID 1094694), through all redshifts. All of the companions' mass ratios are below 1:10 with most even below 1:1000.

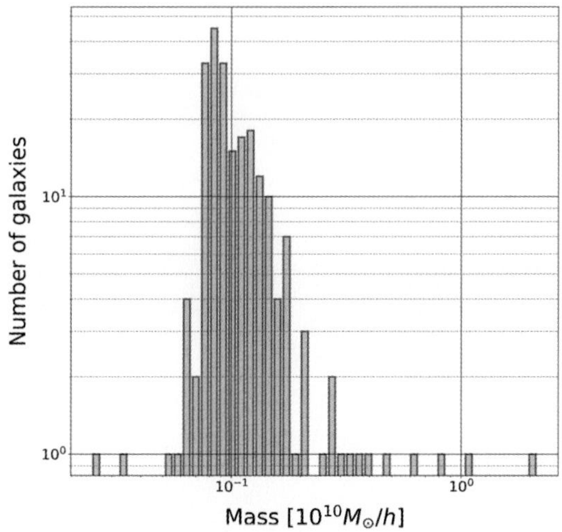

Figure 4. Mass distribution of the merger companions of our selected large, quiescent galaxy (ID 1094694), through all redshifts. More than 30% of the companions have a mass $\log(M_*/M_{\odot}) \geq 9$

References

Hwang, Ho Seong *et al.* 2019, *MNRAS*, 489, 339
Mihos, J. Christopher *et al.* 1996, *ApJ*, 464, 641
Moustakas, John *et al.* 2013, *ApJ*, 767, 50
Pontzen, Andrew *et al.* 2017, *MNRAS*, 465, 547
Pearson, William J *et al.* 2019, *A&A*, 631, A51
Gardner, Jonathan P. *et al.* 2006, *Space Sci. Rev.*, 123, 485
Kalirai, Jason *et al.* 2018, *Contemporary Physics*, 59, 251
Pillepich, Annalisa *et al.* 2018, *MNRAS*, 475, 648
Marinacci, Federico *et al.* 2018, *MNRAS*, 480, 5113
Naiman, Jill P *et al.* 2018, *MNRAS*, 477, 1206

Nelson, Dylan *et al.* 2018, *MNRAS*, 475, 624
Springel, Volker *et al.* 2018, *MNRAS*, 475, 676
Koncz, Bendegúz *et al.* 2023, *Contrib. of the Astron. Obs. Skalnaté Pleso*, 53, 153
Koncz, Bendegúz *et al.*, *in prep.*
Kim, Seong Jin *et al.* 2023, *MNRAS*, 527, 5525
Madau, Piero *et al.* 2014, *Annu. Rev. Astron. Astrophys.*, 52, 415
Suess, Katherine A. *et al.* 2023, *AAS*, 956, L42

The First Chapters of Our Cosmic History with JWST
Proceedings IAU Symposium No. 391, 2025
T. Nanayakkara & M. Maseda, eds.
doi:10.1017/S1743921325000055

Unveiling the ionizing sources in galaxies with flexible nebular emission modeling

Yijia Li[1], Joel Leja[1], Ben Johnson[2], and Blue Jay team

[1]The Pennsylvania State University, University Park, PA 16802, USA

[2]Center for Astrophysics | Harvard & Smithsonian, Cambridge, MA, USA
email: `yjli224444@gmail.com`

Abstract. What reionized the universe? What are the masses, star formation rates, and formation histories of galaxies in the early universe? Answering these questions requires a detailed understanding of the nebular emission properties in galaxies. However, the complex physics governing nebular emission in galaxies, particularly in the early universe, often defy simple low-dimensional models. We present `Cue`, a highly flexible tool for interpreting nebular emission across a wide range of abundances and ionizing conditions of galaxies at different redshifts. Unlike typical nebular models used to interpret extragalactic nebular emission, our model does not require a specific ionizing spectrum as a source, instead approximating the ionizing spectrum with a 4-part piece-wise power-law, along with freedom in C/O, N/O, O/H, electron density, and total ionizing photon budget. This flexibility allows us to either marginalize over or directly measure the incident ionizing radiation, thereby directly interrogating the source of the ionizing photons in distant galaxies via their nebular emission. We demonstrate the use of Cue on a variety of simulated and real JWST data. For example, Cue reproduces the emission lines and ionizing production efficiency of GS-NDG-9422, a peculiar $z = 6$ galaxy with a claimed top-heavy IMF, with an more orthodox ionizing continuum consistent with a combination of normal low-metallicity stellar populations and a low-luminosity AGN. Additionally, we show the inferences of Cue applied to JWST/NIRSpec spectroscopy of $z \sim 2$ galaxies (PI: Belli, ID 1810), specifically identifying the dominant ionizing sources in normal galaxies at the cosmic noon.

Keywords. Photoionization; Stellar populations; Spectral energy distribution; Galaxy evolution

---

1. `Cue`: a fast and flexible nebular emulator

Nebular emission in galaxies probes the chemical abundances, density, and ionization states of the gas around the ionizing source. It plays an important role in properly characterizing the properties of these sources such as star formation rates (SFRs) or active galactic nucleus (AGN) activities, and constrains the ionizing budget of galaxies. With the advanced spectroscopic capabilities of *JWST* extending beyond cosmic noon, we can now chart the properties of multi-phase gas, particularly ionized gas, in unprecedented detail across cosmic time. This progress renews the importance of a long-standing question in observations of distant galaxies: what sources are ionizing the gas in these galaxies, and how can we learn about their properties? However, the complexity of nebular physics poses challenges in modeling galaxies in the early universe with unexpected nebular emission (e.g., Bunker et al. 2023; Curti et al. 2023; Abdurro'uf et al. 2024; Cameron et al. 2024; Carniani et al. 2024; Zavala et al. 2024)) especially when the sources powering the emission are unknown.

Moreover, the accuracy of nebular emission modeling has profound effects on the interpretation of the photometry and spectra of high-redshift galaxies, as nebular emission

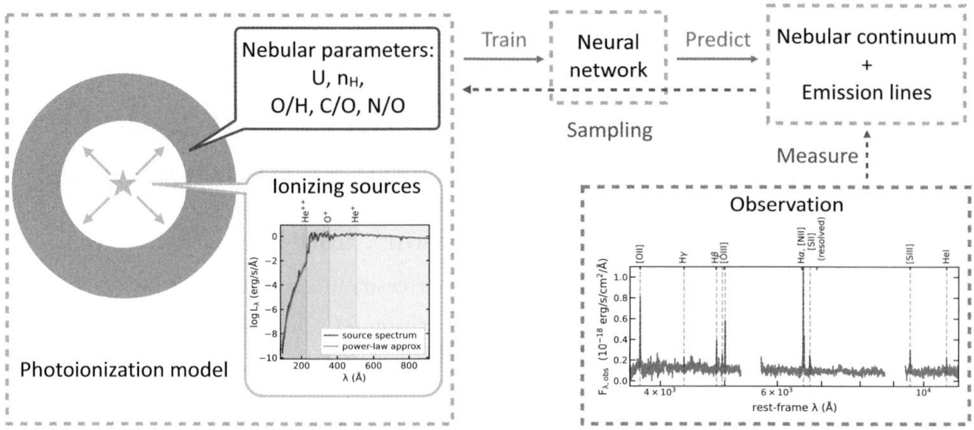

Figure 1. A schematic diagram demonstrating the main inputs and outputs of `Cue`, and how this neural net (NN) emulator is applied to modeling nebular emission observations by sampling the distribution of nebular and ionizing spectrum parameters. In total, `Cue` has 5 free nebular parameters, U, $n_{\rm H}$, [O/H], [N/O], and [C/O], and 7 free parameters characterizing the shape of the ionizing spectrum. The emulator consists of 1 NN for nebular continuum and 14 NNs for emission line prediction.

contributes $\gtrsim$20% of UV and optical emission of star-forming galaxies (e.g., Reines et al. 2010; Stark et al. 2013; Pacifici et al. 2015), with increasing importance towards the high specific star formation rates (sSFRs) measured at high redshifts. The uncertainties associated with the nebular model will not only affect the inferred ionizing gas properties but also influence all the estimated galaxy properties including mass, SFR, dust properties, etc. While previous nebular emission studies are often based on galaxies at $z < 3.5$, high-redshift galaxies exhibit different nebular conditions and ionizing sources, which may suggest lower metallicity, α-enhanced stellar populations with harder ionizing spectra in these galaxies (e.g., Cameron et al. 2023; Sanders et al. 2023; Nakajima et al. 2023; Tacchella et al. 2023, Shapley et al. 2024). Therefore, the applicability of the nebular models needs to be extended to galaxies of a wider redshift range.

To address this, we developed a highly flexible nebular emulator for modeling the continuum and line emission from individual HII regions, named `Cue` (Li et al. 2024a). By leveraging neural networks to emulate the spectral synthetic code `CLOUDY`, `Cue` covers an extensive parameter space, making it suitable for addressing challenges posed by the unique chemistry and ionizing properties of galaxies in the early universe (Figure 1). It models the input ionizing spectra at $1\,\text{Å} \leq \lambda \leq 912\,\text{Å}$ as a flexible 4-part piecewise-continuous power-law, along with freedom in gas density $n_{\rm H}$, ionization parameter U, [O/H], [C/O], and [N/O], enabling self-consistent nebular emission predictions for specific stellar models like one with a top-heavy IMF. Such flexibility allows direct investigation of the ionizing source through emission lines by marginalizing over the incident ionizing radiation and the chemical conditions. By taking advantage of the compact neural networks, `Cue` achieves a high accuracy at the level of 5% for nebular continuum and 128 emission lines from UV to submillimeter and a fast execution time of $\lesssim$6 ms per prediction.

One main feature of `Cue` is its source agnosticism. Unlike typical nebular models used in galaxy SED-fitting, `Cue` no longer relies on a set of stellar isochrones and spectral libraries, but instead can take in piece-wise power-law ionizing spectra of a wide range. We design the allowed range of our ionizing spectrum parameterization to incorporate various types of astrophysical sources shown in Figure 2, including young ($t_{\rm age} < 25\,$Myr) simple stellar

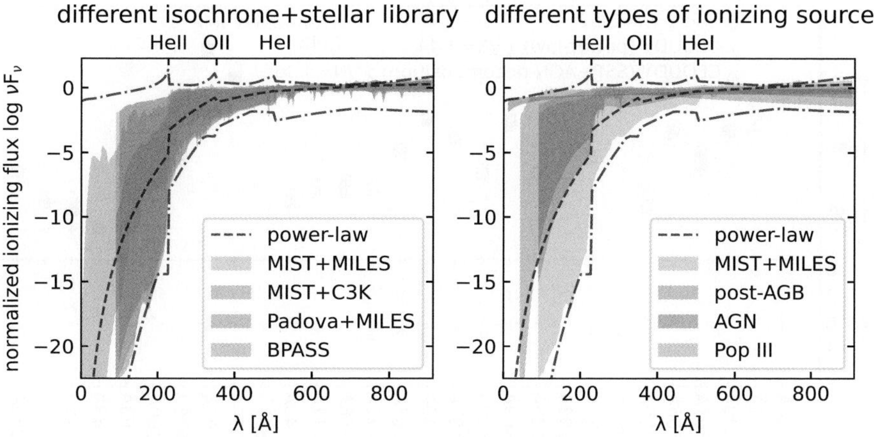

Figure 2. The range of ionizing spectra from different sources that can be modeled using the training set. The left panel compares the range of ionizing spectra over many stellar ages and metallicities, in different stellar isochrones and spectral libraries. The right panel compares the range of ionizing spectra from different types of sources. The black dashed line represents the median of our 4-segment power-law ionizing spectrum in the training set, and the lower and upper limits of the training ionizing spectra are in dashed-dotted lines. The ionizing spectrum shape provides insights into the type of ionizing sources and the stellar isochrones and libraries and `Cue` describes the input ionizing spectrum with a flexible 4-part continuous power-law to encompass various types of ionizing spectra. Adapted from Li et al. (2024a).

populations (SSPs), post-asymptotic giant branch (post-AGB) stars, PopIII stars, and AGNs. Our mock tests suggest that `Cue` accurately capture the shape of the ionizing spectra and the nebular properties based on UV–NIR emission lines at $S/N(\mathrm{H}_\alpha) \gtrsim 10$, and furthermore capable of distinguishing between the ionizing spectra predicted by single and binary stellar models.

2. Modeling the ionizing radiation of GS9422 with `Cue`

Recently, Cameron et al. (2024) argued that a galaxy at $z = 5.943$, JADES-GS+53.12175−27.79763 (hereafter GS9422), has nebular lines and UV continuum emission that appears to require a high ionizing photon production efficiency. This has been explained with an exotic stellar initial mass function (IMF), 10-30x more top-heavy than a Salpeter IMF. Cameron et al. (2024) highlights two main observational constraints that disfavor typical ionizing sources; a UV turnover at $\lambda \approx 1430$ Å combined with a Balmer jump suggest strong two-photon nebular continuum emission and high neutral hydrogen absorption for Lyα; and the detected HeII emission lines are faint for AGN, or X-ray binaries. However, as a top-heavy IMF would represent an extraordinary and unprecedented scenario, we explore if there is an alternate explanation for the observed spectral features in Li et al. (2024b).

We model the emission line fluxes of GS9422 with `Cue`, finding a high ionization parameter $\log U = -1.96$, a gas density $\log n_\mathrm{H} = 3.19\,\mathrm{cm}^{-3}$, $\log[\mathrm{O/H}] = -1.01$, $\log[\mathrm{N/O}] = -0.61$, $\log[\mathrm{C/O}] = 0.03$. `Cue` is able to achieve a good fit the the observed lines and UV continuum at 1500 Åusing a double power-law ionizing spectrum which we can decompose in a straightforward fashion into a combination of low-metallicity young stellar population and an AGN in a 65%/35% ratio (Figure 3). In our combined SSP and AGN model the observed nebular continuum is only 24% of the UV flux. This suggests a significantly lower nebular continuum contribution to the observed UV flux (24%) than a top-heavy IMF ($\gtrsim 80\%$), and hence, necessitates a damped Lyman-α absorber (DLA)

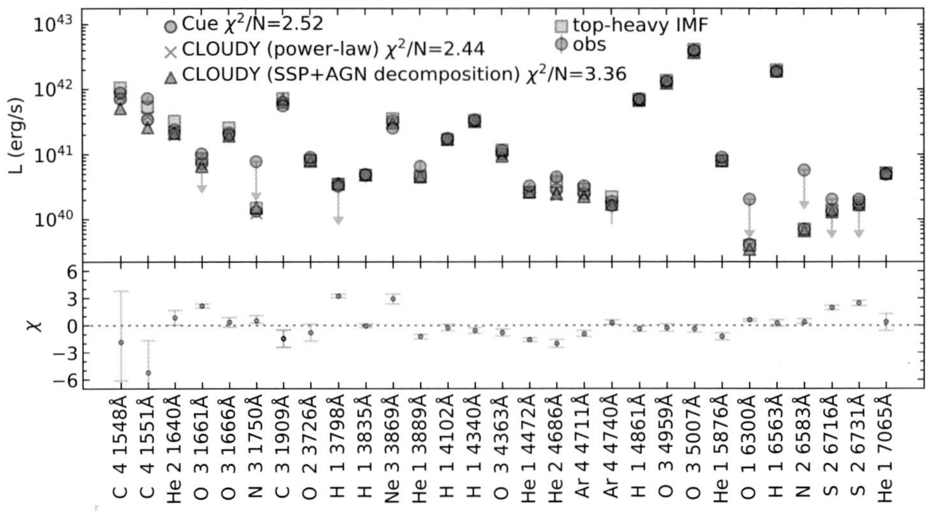

Figure 3. Emission line fit results of GS9422. Upper panel: emission line measurements along with their uncertainties from Cameron et al. (2024) are in blue. The `Cue` posterior median assuming a double power-law ionizing spectrum and five free nebular parameters is in orange. To confirm the `Cue` result, we also run `CLOUDY` using the ionizing spectrum and nebular parameters `Cue` predicted as input, and the output line luminosities are shown in green. Additionally `CLOUDY` prediction from an SSP ($t_{\rm age} = 3.2\,{\rm Myr}$ and $\log(Z/Z_\odot) = -1.34$) with the proposed top-heavy IMF is in grey. The `CLOUDY` output given a combination of a MIST+C3K SSP and an AGN ionizing spectrum as the input is in purple. Lower panel: the residual of the `Cue` fit results denominated by the measurement uncertainties. The error bars indicate the 1σ posterior range of χ. The results of `Cue` fit and the SSP+AGN fit match the observations reasonably well. Adapted from Li et al. (2024b).

with high neutral hydrogen column density $\log N(\mathrm{HI}) \gtrsim 22.7\,\mathrm{cm}^{-2}$ to explain the continuum turnover bluewards of ~ 1400 Å. While current data cannot rule out either scenario: a mixed stellar and AGN ionizing source, or a remarkably top-heavy IMF, given the immense impact the proposed top-heavy IMF would have on models of galaxy formation, it is important to propose viable alternative explanations and to further investigate the nature of peculiar high-z nebular emitters.

3. Applying `Cue` to low-mass star-forming galaxies at cosmic noon

We apply `Cue` to ~46 hours of *JWST* NIRSpec spectroscopy of $z \sim 2$ galaxies from the Cycle 1 program Blue Jay (PI: Belli, ID 1810). Comparing the emission-line modeling results from `Cue` with the spectral energy distribution (SED) modeling results from `Prospector` (Johnson et al. 2021), we find a large scatter in the ionizing photon budget (Q) at fixed SFR. This suggests significant systematic offsets in emission line-based SFRs, likely driven by variations in stellar metallicity, star-formation history, binary star fraction, and other stellar population properties.

As illustrated in Figure 4, adopting the widely used SFR-Q relationship produces $\gtrsim 0.3\,\mathrm{dex}$ systematic uncertainty in the Hα-based SFR at $z \sim 2$, with Q potentially spanning a factor of 20 at fixed SFR. This limits our ability to constrain the SFH of the distant galaxies until we can accurately determine the ionizing budgets of their stellar populations. Moreover, >50% of the low-mass dust-free sub-sample show ionizing radiation harder than that typically produced by AGN, motivating further exploration of extreme stellar models.

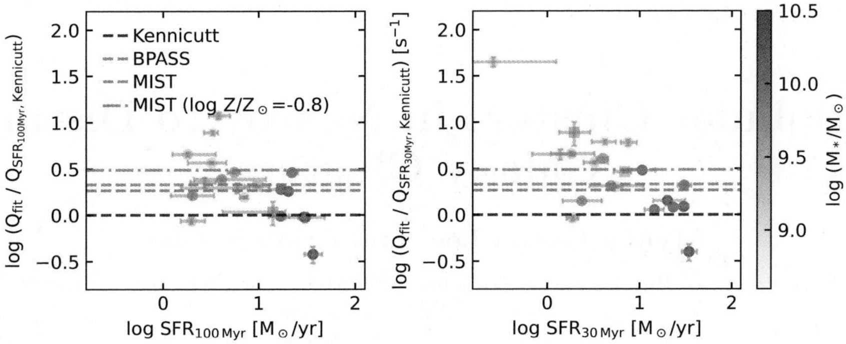

Figure 4. Comparison of Q inferred from `Cue` fits (red) and the standard Kennicutt (1998) estimates (black) which assume a linear SFR-Q relationship, in a sample of low-mass, dust-free star-forming galaxies (SFGs) at $z \sim 2$ from the Blue Jay survey. The left and right columns show SFR averaged over the most recent 100 Myr and 30 Myr, respectively. The results reveal substantial systematic uncertainties in the standard approach. Stellar models that include binary evolution or stellar rotation, such as MIST and BPASS produce more ionizing photons than predicted by Kennicutt (1998) by $\gtrsim 0.3$ dex. Additionally, moving from solar metallicity (dashed lines) to the sample's average stellar metallicity of $\log Z/Z_\odot = -0.8$ (dash-dotted line) increases the predicted ionizing photon production by ~ 0.2 dex.

4. Outlook

`Cue` paves a way for probing the ionizing spectrum of galaxies, which due to neutral gas absorption is only accessible by the nebular emission they power, and thereby interrogating the ionizing source directly. We have demonstrated the power of `Cue` in interpreting different sources at different redshifts using *JWST* data. Going forward, `Cue` will be a useful tool in several contexts – due to its flexibility it can interpret unique sources powered by unusual or mixed ionizing spectra, while also due to its speed it can be applied to industrial-scale spectroscopic surveys. These studies provide insights on the evolution of nebular conditions and ionization sources of galaxies over cosmic times.

References

Abdurro'uf, Larson, R. L., Coe, D., et al. 2024, *ApJ*, 973, 47
Bunker, A. J., Saxena, A., Cameron, A. J., et al. 2023, *A&A*, 677, A88
Cameron, A. J., Katz, H., Witten, C., et al. 2024, *MNRAS*
Cameron, A. J., Saxena, A., Bunker, A. J., et al. 2023, *A&A*, 677, A115
Carniani, S., Hainline, K., D'Eugenio, F., et al. 2024, *Nature*, 633, 318
Curti, M., D'Eugenio, F., Carniani, S., et al. 2023, *MNRAS*, 518, 425
Johnson, B. D., Leja, J., Conroy, C., & Speagle, J. S. 2021, *ApJS*, 254, 22
Kennicutt, Jr., R. C. 1998, *ARA&A*, 36, 189
Li, Y., Leja, J., Johnson, B. D., et al. 2024a, arXiv e-prints, arXiv:2405.04598
Li, Y., Leja, J., Johnson, B. D., Tacchella, S., & Naidu, R. P. 2024b, arXiv e-prints, arXiv:2404.02333
Nakajima, K., Ouchi, M., Isobe, Y., et al. 2023, *ApJS*, 269, 33
Pacifici, C., da Cunha, E., Charlot, S., et al. 2015, *MNRAS*, 447, 786
Reines, A. E., Nidever, D. L., Whelan, D. G., & Johnson, K. E. 2010, *ApJ*, 708, 26
Sanders, R. L., Shapley, A. E., Topping, M. W., Reddy, N. A., & Brammer, G. B. 2023, *ApJ*, 955, 54
Shapley, A. E., Sanders, R. L., Topping, M. W., et al. 2024, arXiv e-prints, arXiv:2407.00157
Stark, D. P., Schenker, M. A., Ellis, R., et al. 2013, *ApJ*, 763, 129
Tacchella, S., Johnson, B. D., Robertson, B. E., et al. 2023, *MNRAS*, 522, 6236
Zavala, J. A., Castellano, M., Akins, H. B., et al. 2024, *Nature Astronomy*

The First Chapters of Our Cosmic History with JWST
Proceedings IAU Symposium No. 391, 2025
T. Nanayakkara & M. Maseda, eds.
doi:10.1017/S1743921325001620

Globular Clusters in Nearby to Distant Galaxy Clusters

Myung Gyoon Lee[1] **and Brian S. Cho**[2]

Department of Physics and Astronomy, SNUARC, Seoul National University, Seoul, Republic of Korea
email: [1]`mglee@snu.ac.kr`, [2]`brianscho@snu.ac.kr`

Abstract. Globular clusters in the nearby universe were formed in the first chapter of our cosmic history, but they show only the current chapter. However, the JWST enables us to see the younger phase of the globular clusters in the distant universe. Galaxy clusters host both galaxy globular clusters and intracluster globular clusters. We present a wide field photometric survey of globular clusters in the nearby to distant galaxy clusters. We discuss the photometric properties of these globular clusters in relation with the structure and assembly history of their host galaxy clusters.

Keywords. Globular Clusters, Intracluster Globular Clusters, Early-type Galaxies, Galaxy Clusters

---

1. Introduction & Data

Globular clusters (GCs) are among the oldest objects today. There are two types of GCs that are identified based on their location. The first is galaxy globular clusters that belong to their host galaxy, and the second is intracluster globular clusters (IGCs) that reside outside of galaxy regions and belong to their host galaxy cluster. IGCs are thought to be tidally stripped from the outskirts of individual galaxies during galaxy interactions. Thus, IGCs hold vital information about the galaxy they originated from and the dynamical evolution of their host galaxy cluster.

IGCs have been detected in several nearby galaxy clusters, but previous studies have mostly focused on the cluster center regions due to the limited availability of deep and wide-field data. However, the Subaru/Hyper Suprime-Cam (HSC) offers a large field of view and depth that allows GC detection past the GC luminosity function turnover magnitude for nearby clusters.

In this study, we use archival HSC g and r-band data to identify GCs and use them as a probe to investigate the spatial extent and assembly history of three nearby galaxy clusters in different dynamical stages: Hydra I (Abell 1060), Coma (A1689), and Leo (A1367). We then compare our results with JWST/NIRcam results for a distant galaxy cluster, SMACS J0723.3-7327, located at a redshift of z = 0.39.

2. Results & Discussion

We select GC candidates based on color $(0.35 < (g-r)_0 < 1.00)$, and magnitude $(-11 < M_r < -7.5$ for Hydra I, $-12.5 < M_r < -8.5$ for Coma and Leo). We divide the GC population into three subgroups: all, blue, and red GCs with a color cut at $(g-r)_0 = 0.655$.

Figure 1 shows the large-scale spatial distribution of globular clusters and bright red-sequence (RS) galaxies. In each cluster, the strongest concentration of GCs is at the

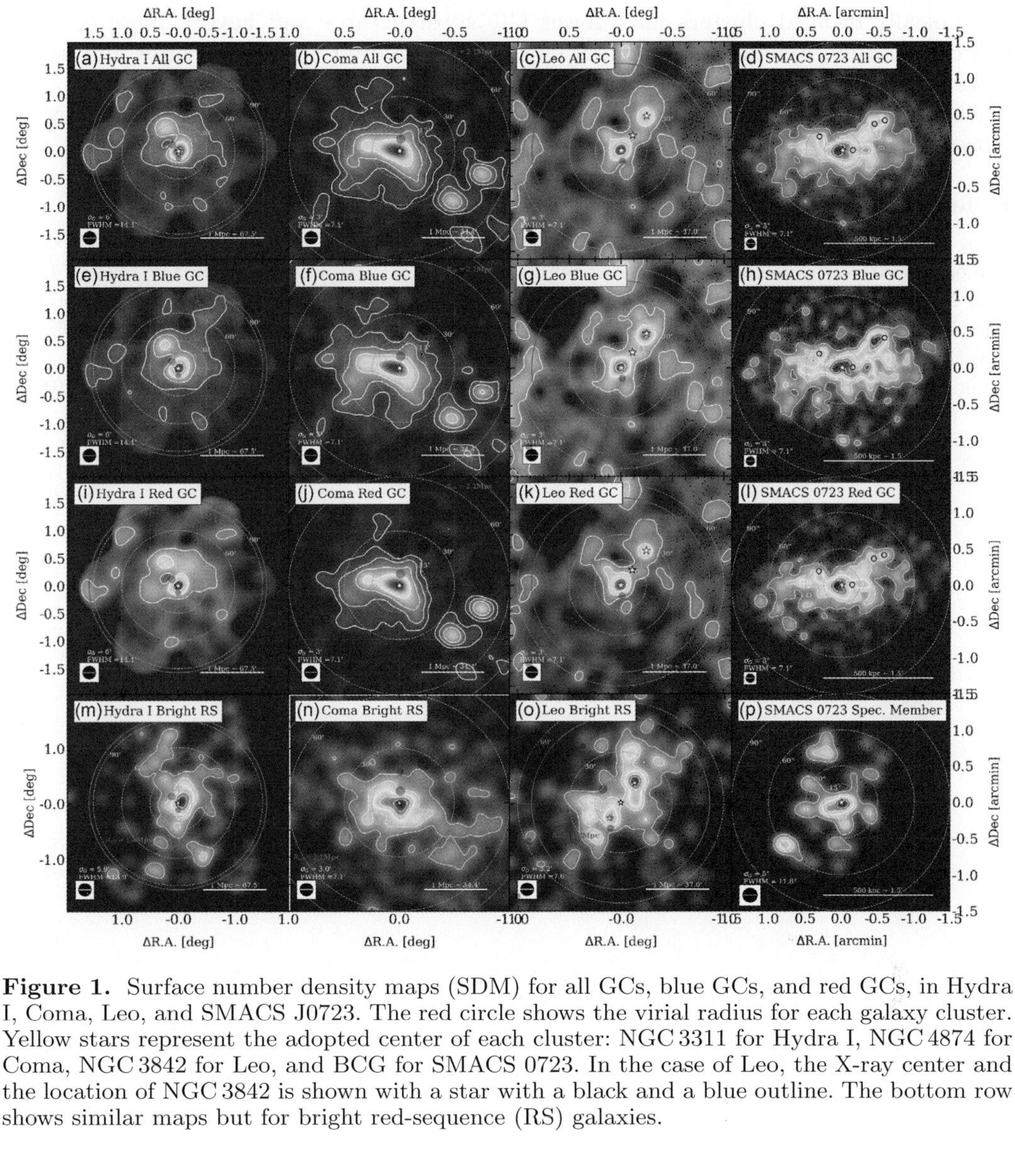

Figure 1. Surface number density maps (SDM) for all GCs, blue GCs, and red GCs, in Hydra I, Coma, Leo, and SMACS J0723. The red circle shows the virial radius for each galaxy cluster. Yellow stars represent the adopted center of each cluster: NGC 3311 for Hydra I, NGC 4874 for Coma, NGC 3842 for Leo, and BCG for SMACS 0723. In the case of Leo, the X-ray center and the location of NGC 3842 is shown with a star with a black and a blue outline. The bottom row shows similar maps but for bright red-sequence (RS) galaxies.

cluster center and there exists a large scale IGC component that is dominated by blue GCs. In Hydra I, the GCs show two strong overdensities: at the cluster center and at the galaxy group of IC 2597. The blue GCs also show an IGC component that extends between these two structures, suggesting that these two structures interacted in the past. In Coma, the GCs show multiple overdensities: at the positions at the cluster center and brightest galaxies. The overall distribution is asymmetric and extends from SW to NE. In Leo, the GCs show several overdensities, and also some outside of the virial radius. The strongest GC overdensity is at NGC 3862 and NGC 3842. There is also a structure of GCs that extends between the two galaxies. Finally, the GCs of SMACS 0723 show an asymmetric distribution with several substructures.

Overall, the morphology of the IGC structures varies from cluster to cluster, and it indicates that the clusters are in different states of dynamical evolution. Furthermore, the IGC substructures indicate the past interaction between galaxies. We find that dynamically evolved clusters have fewer GC substructures and have a more compact GC distribution, dynamically younger clusters have more GC substructures and have a more asymmetric and elongated GC distribution.

This work was supported by the National Research Foundation grant funded by the Korean Government (RS-2024-00340832).

References

Cho, B. S., Lee, M. G., & Jang, I. S. et al. 2024 in prep.

Hilker, M. 2015, Proceedings IAU Symposium No. 317, 2015. doi:10.48550/arXiv.1512.07417

Lee, M. G., Park, H. S., & Hwang, H. S. 2010, *Science*, 328, 334. doi:10.1126/science.1186496

Lee, M. G., Bae, J. H., & Jang, I. S. 2022, *ApJL*, 940, L19. doi:10.3847/2041-8213/ac990b

Oh, S.-A., Lee, M. G., & Jang, I. S. 2023, *ApJ*, 944, 51. doi:10.3847/1538-4357/acb1b1

Peng, E. W., Ferguson, H. C., Goudfrooij, P., et al. 2011, *ApJ*, 730, 23. doi:10.1088/0004-637X/730/1/23

The First Chapters of Our Cosmic History with JWST
Proceedings IAU Symposium No. 391, 2025
T. Nanayakkara & M. Maseda, eds.
doi:10.1017/S1743921324003326

Temporal structure in the HERA observations that pose a challenge to observing light from the early universe

Mekuanint K. Hailemariam[1], Lerothodi L. Leeuw[1], Miguel F. Morales[2] and David R. DeBoer[3]

[1]Department of Physics and Astronomy, University of the Western Cape, Cape Town, 7535, South Africa

[2]Department of Physics, University of Washington, Seattle, WA

[3]Department of Astronomy, University of California, Berkeley, CA
email: mekunet@gmail.com

Abstract. The Hydrogen Epoch of Reionization Array (HERA) is a staged experiment designed to characterize our cosmic dawn using the redshifted 21 cm line of neutral hydrogen from the epoch of reionization. The bright extragalactic foreground emission dominates the extremely faint 21-cm line signal from the epoch of reionization. Along with the foreground emission, the antenna systematics make it challenging to accurately measure the extremely weak signal. The systematics result in high or low power, abnormal bandpass shape, excess correlation, and temporal structure. We examine the temporal discontinuity issues, which are the main source of temporal structure.

Keywords. HERA systematics, temporal discontinuity, temporal variability, temporal structure

1. Introduction

HERA is dedicated to observing large-scale structures during and prior to the Epoch of Reionization (EoR). The full HERA instrument is a 350-element interferometer consisting of 14 m parabolic dishes observing from 50 to 250 MHz. HERA's antenna element and its compact configuration are optimized for robust power spectrum detection (DeBoer *et al.* (2017)) from the EoR. HERA's primary objective is to characterize the less explored cosmic dawn, which includes the period from the formation of the first stars and black holes to the full reionization of the intergalactic medium (Shimabukuro *et al.* (2022)). In comparison to the foreground emissions, the 21 cm signal from the EoR is extremely weak (Santos (2005), Bernardi *et al.* (2010)). Furthermore, systematics that arise along the signal chain in the HERA antenna system could easily alter the signal, making it challenging to observe the 21 cm emission. This brief analysis is part of the effort to characterize and identify origins of temporal structure. The comprehensive analysis of this study is being conducted and will be available soon.

2. Temporal Structure

Excess correlation, high/low power, bad bandpass shape, and temporal structure are the well-known failure modes in the HERA observations. The work of Storer *et al.* (2022) provides a detailed explanation of how to identify antenna malfunctions associated with the failure modes listed above. The paper discussed two crucial metrics-temporal

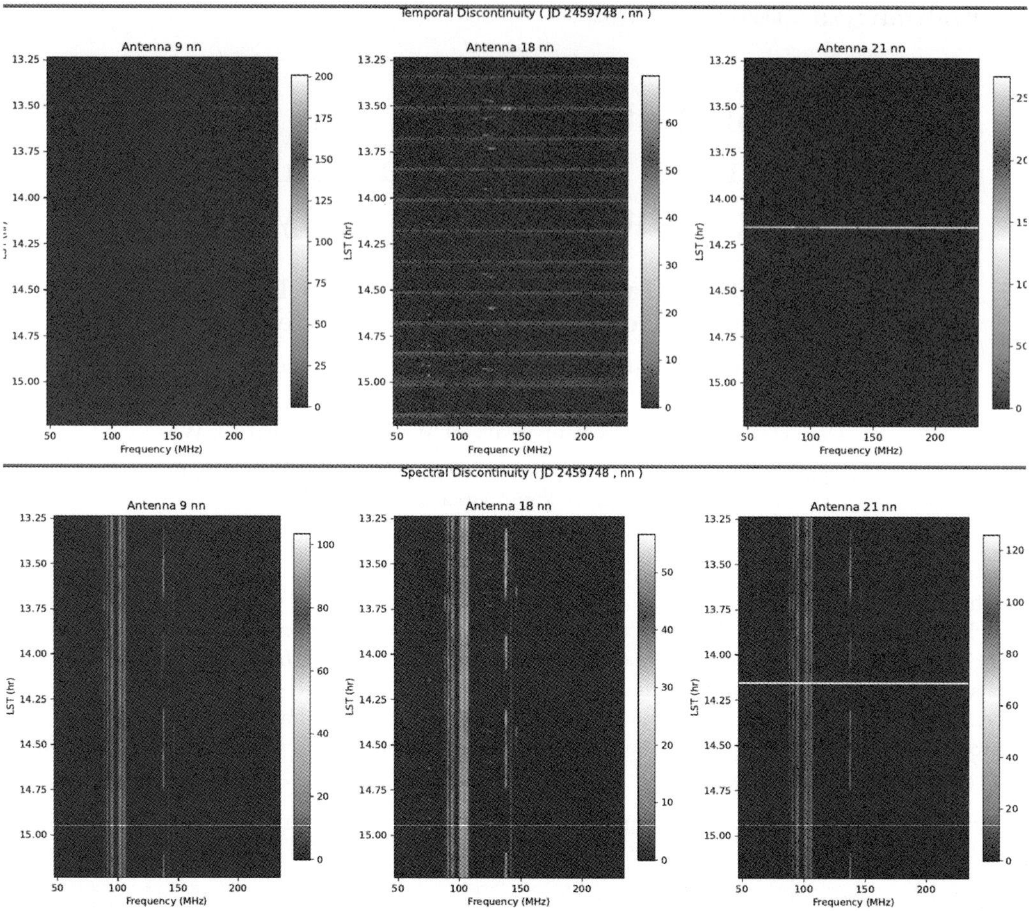

Figure 1. Temporal discontinuity index (top panels) and the spectral discontinuity index (bottom panels). We pick antenna 9 as a reference because it has low modified zscores for the temporal variability and temporal discontinuity metrics across all of the data we looked at. All of the antennas affected by the first type of systematics (antenna 18) show excess RFI in addition to the temporal discontinuity issue. This could affect the band pass shape. Furthermore, it is found that the temporal discontinuity seen in antenna 21 is accompanied by low power.

discontinuity and temporal variability metrics-for detecting problems related to temporal structure issues. To gain a better understanding of the causes of temporal discontinuity issues, we propose a temporal and spectral discontinuity index to be used in conjunction with the temporal variability and temporal discontinuity metrics. We define the temporal discontinuity index for antenna i and polarization p as

$$TDI(t,\nu) = \frac{V_{ii,pp}(t+\Delta t,\nu) - V_{ii,pp}(t,\nu)}{\langle V_{ii,pp}(t+\Delta t,\nu) - V_{ii,pp}(t,\nu)\rangle_t} \tag{1}$$

and the spectral discontinuity index as

$$SDI(t,\nu) = \left| \frac{V_{ii,pp}(t,\nu+\Delta\nu) - V_{ii,pp}(t,\nu)}{\langle V_{ii,pp}(t,\nu+\Delta\nu) - V_{ii,pp}(t,\nu)\rangle_t} \right| \tag{2}$$

where $V_{ii,pp}(t,\nu)$ is the auto-correlation of the visibility. After selecting problematic antennas using the Pearson correlation of the temporal discontinuity index, we plotted the spectral and temporal discontinuity index for the identified antennas, as shown in Figure 1.

3. Summary and conclusions

Multiple temporal discontinuity issues have been identified. A single antenna may be affected by one or more types of systematics. The same kind of sytematics that produce the temporal discontinuity may also affect bandpass shape and/or power.

References

Bernardi G. and de Bruyn A. G. and Harker, G., *et al.* 2010, *A&A*, 522, A67
DeBoer David R., Parsons Aron R., Aguirre James E., *et al.* 2017, *PASP*, 129, 974, 045001
Santos M'ario G. and Cooray Asantha and Knox Lloyd, 2005, *ApJ*, 625, 2, 575-587
Shimabukuro Hayato, Hasegawa Kenji, Kuchinomachi Akira., *et al.* 2022, *PASJ*
Storer Dara and Dillon Joshua S. and Jacobs Daniel C., *et al.* 2022, *Radio Science*, 57, 1, e07376

Author index